Valuation of Intellectual Property and Intangible Assets

Third Edition

2004 Cumulative Supplement

Gordon V. Smith
Russell L. Parr

John Wiley & Sons, Inc.

Published by John Wiley & Sons, Inc., Hoboken, New Jersey.
Published simultaneously in Canada.

Library of Congress Cataloging-in-Publication Data
Smith, Gordon V., 1937—
Valuation of intellectual property and intangible assets / Gordon V. Smith, Russell L. Parr.—3rd ed.
p. cm.—(Intellectual property series (John Wiley & Sons))
Includes bibliographical references and index.
ISBN 0-471-36281-6 (cloth : alk. paper)
0-471-46469-4 (Supplement)
1. Intellectual property—Valuation—United States. 2. Intangible property—Valuation—United States. I. Parr, Russell L. II. Title. III. Series.
KF2979.S65 2000
346.7304'8—dc21 99-053661

Printed in the United States of America.

10 9 8 7 6 5 4 3 2 1

Support, love, and encouragement are still
the most valuable elements of life.
Dorothy Parr has provided a limitless supply of all. . . .
Thanks again, Mom!

To some of the pioneers in the appraisal profession,

John L. Moon, William M. Young,
F. S. Olson, A. B. Hossack

To some others who carried on their beginning,

Clement J. Schwingle,
Wayne D. Georgeson, G. V. Giese,
C. E. O. "Johnny" Walker, Ettore Barbatelli,
John B. Hossack

And to some fondly remembered mentors of that generation,

"Doc" Heimbach and "Hal" Britton

ABOUT THE AUTHORS

Gordon V. Smith is President of AUS Consultants. A graduate of Harvard University, he has been in the valuation profession for over 35 years and has served a broad spectrum of international clients. He is an Adjunct Professor of Law at Franklin Pierce Law Center and has lectured across the United States and in China, Korea, Europe, and South America. He is the Founder and Chairman of the Board of the Intellectual Property Management Institute. An active member of the International Trademark Association, Mr. Smith is the author of *Trademark Valuation* and *Corporate Valuation: A Business and Professional Guide* and co-author of *Intellectual Property: Licensing and Joint Venture Profit Strategies,* all published by John Wiley & Sons. His writings also include contributions to other Wiley intellectual property books.

Russell L. Parr, CFA, ASA, is President of Intellectual Property Research Associates and internationally known for his expertise in intellectual property royalty rates. He holds a Master's in Business Administration and a B.S. in Electrical Engineering from Rutgers University. Mr. Parr is responsible for completion of complex consulting assignments involving the valuation and pricing of patents, trademarks, copyrights, and other intangible assets. His opinions are used to accomplish licensing transactions, mergers, acquisitions, transfer pricing, litigation support, collateral-based financing, and joint ventures. He advises banks about the use of intangible assets as loan collateral and has served as an expert witness regarding intellectual property infringement damages. Mr. Parr is founder of the highly respected *Licensing Economics Review*, which is dedicated to reporting detailed information about the economic aspects of intellectual property licensing and joint venturing. His professional designations include Chartered Financial Analyst from the Association for Investment Management and Research and Accredited Senior Appraiser of the American Society of Appraisers.

PREFACE

Intellectual property and intangible assets continue to drive the world's business. When we wrote the first edition in 1989, we said that intangible assets and intellectual property were "coming of age." In the second edition we noted their "arrival," and the third edition chronicles their continued growth.

Evidence of that growth is at everyone's fingertips. Nearly everyone has access to the Internet through a keyboard at home or at work. E-commerce has exploded and the world is available online, offering nearly every type of product or service imaginable, from banking to securities trading to auctions. We watched the stock market as Internet-related companies attracted billions in capital. Those billions were not to purchase land and build factories but rather to continue the creation and growth of the intangible assets and intellectual property that gave rise to these enterprises in the first place.

We then witnessed the collapse of many e-commerce business models, and some of the invested billions have become millions and some have become vapor. The contraction in the market is merely an affirmation that the foundation of value for intangible assets and intellectual property is *return on investment.* No one is predicting that Internet-based business will go away, any more than anyone is denying the existence and importance of the intangible properties that drove their creation. This is a business world of continual change, and we will continue to attempt to analyze that change through the lens of valuation theory.

Chapter 4A discusses a major change in the accounting principles relating to intangible assets and intellectual property. Statement of Financial Accounting Standards Nos. 141 and 142, issued in June 2001, supersede APB 16 and 17 that we presented in Chapter 4. As a result of these changes, intangible assets acquired by reporting corporations after July 1, 2001, will be reflected on the books of the acquiring company in much greater detail. Examination of their useful economic life will be rigorous, and the recorded value of intangibles such as trademarks and goodwill will be subject to annual review. We discuss the thinking not only of the Financial Accounting Standards Board but also the many corporate and professional respondents to the Board's Exposure Drafts. It is interesting to observe the tension between business practicalities and the desire to provide as much information as possible about these assets that are so important in today's business.

Chapter 11A provides a discussion of the important *Daubert* and *Kumho* cases heard in the U.S. Supreme Court. Expert witnesses in the fields of valuation and economics may well be faced with challenges to the admissibility of their testimony, based on how well it measures up to the application of the "scientific method." Professionals will benefit from the juxtaposition of the *Daubert* principles and the traditional methods for developing valuation and damages opinions.

In Chapter 14A, we focus that lens on the Federal Trademark Dilution Act of 1995 (FTDA). Effective January 16, 1996, the FTDA offers to owners of "famous trademarks" the possibility of injunctive relief if someone else uses a trademark in commerce that "causes dilution of the distinctive quality of the famous mark."

The enforcement of this act has vexed courts and litigants alike, as they sought a clear interpretation of what trademark dilution really is and how to detect when it has occurred. Our suggestion is that, even though awards of monetary damages may not be provided for in the law, the economic principles used to quantify economic damages can be useful in solving the conundrums posed by the FTDA.

In its March, 2003 decision in the case of *Moseley v. V Secret Catalogue*, the U.S. Supreme Court fell short of resolving the dilution issues we discussed. It appears that these issues will be around for some time to come.

In Chapter 17A, we examine the relatively new phenomenon of naming rights transactions and the apparent rise of a seemingly new asset connected with tangible property. We have, in recent years, recognized that a stadium, arena, or other public facility has a marketable attribute as a "billboard." Corporations are willing to pay considerable sums to place their identity on such a prominent property. What happens when the original purchaser of such naming rights cannot pay the rent or is acquired by another that has little interest in the naming rights? What if the owner of the facility is in a quandary about a fair price? Valuation issues will arise.

Appendix F presents a list of valuation resources for the use of practitioners. Information about intellectual property licensing transactions is very difficult to obtain, and we discuss some sources we have utilized. Financial discount rates are an obscure subject for many, and we give some assistance in that quarter, besides listing books and other references that can be useful. Finally, we present descriptions of some professional societies that focus on various aspects of intellectual property business.

The Intellectual Property Management Institute (IPMI) is the subject of Appendix G. IPM*i* is a not-for-profit organization dedicated to professional development in the field of intellectual property management. Effective management of intellectual property has become an essential skill in business. The mission of IPM*i* is therefore to encourage the development of intellectual property management theory and practice, and to provide recognition for those who, through educational and professional development, have attained professional status. Appendix G describes how members prepare for the Candidate and Certified Intellectual Property Manager examinations through a self-study program.

January 2004

Gordon V. Smith
Moorestown, NJ

Russell L. Parr
Yardley, PA

Update Service

BECOME A SUBSCRIBER!

Did you purchase this product from a bookstore?

If you did, it's important for you to become a subscriber. John Wiley & Sons, Inc., may publish, on a periodic basis, supplements and new editions to reflect the latest changes in the subject matter that you ***need to know*** in order to stay competitive in this ever-changing industry. By contacting the Wiley office nearest you, you'll receive any current update at no additional charge. In addition, you'll receive future updates and revised or related volumes on a 30-day examination review.

If you purchased this product directly from John Wiley & Sons, Inc., we have already recorded your subscription for this update service.

To become a subscriber, please call **1-877-762-2974** or send your name, company name (if applicable), address, and the title of the product to:

mailing address: **Supplement Department**
John Wiley & Sons, Inc.
One Wiley Drive
Somerset, NJ 08875

e-mail: **subscriber@wiley.com**
fax: **1-732-302-2300**
online: **www.wiley.com**

For customers outside the United States, please contact the Wiley office nearest you:

Professional & Reference Division
John Wiley & Sons Canada, Ltd.
22 Worcester Road
Etobicoke, Ontario M9W 1L1
CANADA
Phone: 416-236-4433
Phone: 1-800-567-4797
Fax: 416-236-4447
Email: canada@wiley.com

John Wiley & Sons Australia, Ltd.
33 Park Road
P.O. Box 1226
Milton, Queensland 4064
AUSTRALIA
Phone: 61-7-3859-9755
Fax: 61-7-3859-9715
Email: brisbane@johnwiley.com.au

John Wiley & Sons, Ltd.
The Atrium
Southern Gate, Chichester
West Sussex PO19 8SQ
ENGLAND
Phone: 44-1243-779777
Fax: 44-1243-775878
Email: customer@wiley.co.uk

John Wiley & Sons (Asia) Pte., Ltd.
2 Clementi Loop #02-01
SINGAPORE 129809
Phone: 65-64632400
Fax: 65-64634604/5/6
Customer Service: 65-64604280
Email: enquiry@wiley.com.sg

CONTENTS

Note to the Reader: Sections not in the main bound volume are indicated by "(New)" after the title. Material new to *this* supplement is indicated by an asterisk (*) in the left margin of the contents and throughout the supplement.

4A (New)

New Developments in Accounting for Intangible Assets

In Chapter 4 of the main text, we discuss accounting and tax issues as they relate to the valuation of intangible assets and intellectual property. Chapter 5 discusses the growing disconnect between the market value of a business, its underlying assets, and the amounts shown on its financial statements. At the beginning of Chapter 5 we noted the considerable differences between accounting standards in the United States and other countries, particularly as they relate to the accounting for intangible assets and intellectual property and the accounting treatment of assets acquired in a business combination. Those writings reflected the situation in the mid- to late 1990s.

In June 2001, the Financial Accounting Standards Board (FASB) issued two statements that substantially alter U.S. accounting for intangible assets and intellectual property acquired in business combinations and thereafter:

- Statement of Financial Accounting Standards No. 141—Business Combinations (SFAS 141)
- Statement of Financial Accounting Standards No. 142—Goodwill and Other Intangible Assets (SFAS 142)

In this supplemental chapter, we discuss these significant changes as well as some of the background and theories underlying them. There are several reasons why we believe this discussion is useful to this book:

1. Financial statements are the keystone in the body of information that is necessary in a valuation of intangible assets or intellectual property.
2. It is therefore necessary to have a clear understanding of what financial statements are intended to represent. We also need to understand which, if any, of these accounting entries are useful indicators of value.
3. If we take the opportunity to examine and understand the theoretical basis for asset accounting, as well as some of the arguments raised in some very difficult decision making about this basis, we can learn a lot about the nature of intangible assets and intellectual property and their significance within a business enterprise.

In Chapter 5, we pointed out the fact that there can be a considerable disparity between the value of a business and the value of its underlying assets as they are reflected in financial statements. We also cited the views of others that traditional financial statements had become less useful because information about intangibles was excluded. It was these concerns that added momentum to the consideration of significant changes in accounting for assets in the United States.

During this same period of time, a new business model was emerging. This model exemplified the intangible asset issue in the extreme, and prompted some to proclaim a "New Economy."

IS THERE A NEW ECONOMY?

During the 1970s and 1980s we witnessed the explosive growth of companies in the semiconductor, software, and personal computer segments. These were companies whose intangible assets and intellectual property were central to their earning power. We began to observe an increasing disconnect between the value of these enterprises and the amounts carried on their books. But nowhere is the issue of accounting statement—intangible asset disparity more evident than in the case of the new e-commerce enterprises that more recently sprang into existence. These were the darlings of Wall Street and easily raised hundreds of millions of dollars from eager investors. While we all observed the fact that these businesses enjoyed incredible market values, we also observed their singular lack of ability to produce profits. We also observed that they were doing this with essentially no visible assets. The application of traditional investment theory could not explain this situation. These enterprises seemed to defy the investment law of gravity that requires value to be supported by earnings (or at least the near-term prospect of earnings). Many leapt to the conclusion that there must be a new economy emerging, one in which the old rules did not apply.

This conclusion must be examined in a larger context. Intangible assets and intellectual property have always been an integral part of a business. We only need to look back to the first Industrial Revolution to learn that John Wilkinson, an eminent English machine tool maker, felt that secrecy was a better means of protecting his innovations than patents, which would make them publicly known. He seems to have been extremely successful in this, as his company produced accurately bored cylinders for James Watt's steam engine company for many years. In fact, Watt and his partner, Matthew Boulton, went to great lengths to obtain from Parliament an extension of Watt's condenser patent for almost 20 years, assuring the market supremacy of the Watt engine. We recall Eli Whitney's unsuccessful efforts to enforce his cotton gin patent and the successful efforts of pioneers such as Henry Bessemer, Edison, Westinghouse, and Ford to defend their patent-protected market positions. These business pioneers (and their investors and lenders) certainly realized the importance of their intangible assets, no matter how they were reflected on financial statements.

In 1955, the concept of "automation" was in everyone's mind. It could very well be that some thought of it as a herald of a new economy in which the old investment theories no longer held. Many disagreed, however:

> One of the frustrations of creative engineers and designers is that many of their achievements turn out to be impractical for economic reasons. The field of automation is likely to abound in such examples. A lot of applications will be ruled out by the necessity for fairly stable product designs and the ability of the market to absorb a much higher volume of output. . . . In short, there will be many instances where automation, while technically feasible, just does not pay. We will rarely hear about such cases: we will hear about technological feasibility much more frequently than about the arithmetic of new investment opportunities that cannot be justified.[1]

Then SEC Chairman, Arthur Levitt, commented in a speech to the Economic Club of New York in 1999:

> The dynamic nature of today's capital markets creates issues that increasingly move beyond the bright line of black and white. New in-

[1] George P. Shultz and George B. Baldwin, *Automation: A New Dimension of Old Problems* (Public Affairs Press, Washington, D.C., 1955) as quoted in Paul A. Samuelson, Robert L. Bishop, & John R. Coleman, *Readings in Economics* (McGraw-Hill, 3d ed., 1958).

> dustries, spurred by new services and new technology, are creating new questions and challenges that must be addressed. Today, we are witnessing a broad shift from an industrial economy to a more service based one; shift from bricks and mortar to technology and knowledge.
>
> This has important ramifications for our disclosure and financial reporting models. We have long had a good idea of how to value manufacturing inventory or assess what a factory is worth. But today, the value of R&D invested in a software program, or the value of a user base of an Internet shopping site is a lot harder to quantify. As intangible assets continue to grow in both size and scope, more and more people are questioning whether the true value—and the drivers of that value—is being reflected in a timely manner in publicly available disclosure.

We feel that Mr. Levitt's statement is a better reflection of what has happened and what should be done than the more dramatic proclamations of a "new economy."

There has been an evolutionary change in the proportions of bricks and mortar and intangible assets as driving forces in business. Intangible assets and intellectual property have not suddenly appeared in the last 10 years. This is well described by Damodaran:

> The value of a firm is based on its capacity to generate cash flows and the uncertainty associated with those cash flows. Generally, more profitable firms have been valued more highly than less profitable ones. In the case of new technology firms, though, this proposition seems to have been turned on its head . . . The negative earnings and the presence of intangible assets is used by analysts as a rationale for abandoning traditional valuation models and developing new ways that can be used to justify investing in technology firms . . . This search for new paradigms is misguided . . . The value of a firm is still the present value of the expected cash flows from its assets.[2]

These accounting changes should therefore be viewed not as a result of the emergence of a new economy, but rather in the context of the evolutionary change in the character of U.S. businesses. They are a significant step in re-

[2] Aswath Damodaran, *The Dark Side of Valuation* 11, 12 (Prentice-Hall, Inc., Upper Saddle River, NJ, 2001).

sponse to the long-recognized need to provide more and better financial information to the many constituents of a business enterprise.

INDEPENDENCE ISSUES

Another issue emerged during recent years that had some influence on the new accounting pronouncements. The Securities and Exchange Commission (SEC) was responding to concerns about auditor independence. Many auditing firms, with the Big Five in the van, had developed extensive consulting practices that offered a wide range of services to clients in addition to the audit function. One of the policy goals of the SEC is to protect "the millions of people who invest their savings in our securities markets in reliance on financial statements that are prepared by public companies and other issuers and that, as required by Congress, are audited by independent auditors."[3] In the pursuit of this goal, the SEC's concern about auditor independence stemmed from the possibility that an auditor might be influenced by the fact that significant nonaudit services were being provided to the client, and that this might impair the auditor's independence.

In an agreement between the SEC and the American Institute of Certified Public Accountants (AICPA), an Independence Standards Board (ISB) was formed in 1997 to initiate research and develop standards and solicit public views relative to auditor independence issues. The ISB was disbanded in 2001, but many of its findings were incorporated into the final SEC auditor independence requirements.

In September 1999, the ISB issued a Discussion Memorandum concerning appraisal and valuation services. This was prompted in part by some specific valuation concerns:

> Recently, the SEC Staff has expressed independence concerns regarding auditor valuations of "in-process research and development costs," as part of an auditor-assisted allocation of the purchase price of an acquired business to its individual assets and liabilities. This allocation assistance has historically been permitted, but the significance of the in-process R&D valuations to the financial statements of some companies has caused the Staff to question whether auditors should perform them for audit clients.[4]

[3] Securities and Exchange Commission, Final Rule: Revision of the Commission's Auditor Independence Requirements [hereinafter "SEC Final Rule"], Executive Summary, p. 2.
[4] Independence Standards Board, Discussion Memorandum (DM 99-3)—Appraisal and Valuation Services, paragraph 5.

To illuminate this statement, assume a business acquisition in which the target company is an early stage business or an early stage division or product line of a mature company. One would expect that such an entity would probably have made a significant investment in the development of technology or software, for example, which was intended to form the basis for some new product or service in the future. At the time of acquisition the economic outcome of that investment in research and development (R&D) is largely unknown. It is also reasonable to assume that the acquiring company would agree to a purchase price of the entity that would compensate the existing owners, in whole or in part, for that investment in research and development. In fact, in a high technology or e-commerce business, it would not be at all surprising to discover that a high percentage of the total purchase price was so identified.

Accounting rules specified that the amount of purchase price allocated to in-process R&D was to be immediately expensed by the acquiring company. The result was that this portion of the purchase price did not appear on the balance sheet of the acquiring company and there was no ongoing amortization of that amount as a result.[5] If the research and development turned out to be successful, the acquiring company would have purchased a valuable business with very little investment shown on its own balance sheet.[6]

The valuation of in-process research and development can have a very significant impact on the future financial results of operation of the acquiring company. The concern of the SEC was that when the acquiring company's auditors were performing this valuation, they were, in effect, auditing their own work in a situation where there was considerable impact on future financial statements. This is further brought into focus in the SEC's comments:

> For example, where a company acquires another company with large, on-going in-process research and development projects, the acquiring company will need to decide how much of the purchase price to allocate to those projects. This may affect in turn the amount charged against earnings in the current year as in-process research and development expense, and the amount to be classified as goodwill and amortized against future years' earnings. Any such alloca-

[5]Under previous accounting rules, the amount of purchase price allocated to unidentified intangible assets was lumped together with goodwill and amortized over a period not to exceed 40 years. This amortization reduced reported earnings. Therefore, business managers were considerably motivated to maximize the amount of purchase price allocated to in-process R&D.

[6] Statement of Financial Accounting Standards No. 142, Financial Accounting Standards Board, footnote 8 on page 4: "Statement 2 and Interpretation 4 require amounts assigned to acquired intangible assets that are to be used in a particular research and development project and that *have no alternative future use* to be charged to expense at the acquisition date."

tions later will be reviewed in the course of the audit, leading the firm to audit its own work.[7]

As a result of all this, although auditors can continue to perform valuations for their clients under certain circumstances, the SEC now restricts auditor valuations "where it is reasonably likely that the results of any valuation or appraisal, individually or in the aggregate, would be material to the financial statements, or where the results will be audited by the accountant."[8] This prohibition specifically includes valuations that serve as the basis for allocations of purchase price, which are the focus of SFAS 141 and SFAS 142.

BACKGROUND

Many, many opinions have been voiced about how to get more and better financial information into the hands of lenders and investors. Business people, investors, lenders, the accounting profession, valuation professionals, and academics have been among these constituencies that have sought to be heard. From our reading, it appears that the suggestions generally fall into one of the following concept categories:

1. A whole new financial reporting scheme is required.
2. Financial reporting should be modified so that internally generated intangible assets and intellectual property could be recognized.
3. Leave the financial statements alone but add additional supplemental information that would provide outsiders with some information about the intangible asset value drivers of a business.
4. Leave the financial statements as they are.

One view of the "start over again" approach was expressed by the Canadian Institute of Chartered Accountants as part of its Canadian performance reporting initiative, begun in 1994. One report that emanated from this initiative stated that:

1. "In addition to the pragmatic concerns registered by business executives, a strong theoretical case can be made that the current accounting

[7] SEC Final Rule, p. 26.
[8] Securities and Exchange Commission, Current Accounting and Disclosure Issues, Division of Corporation Finance, August 31, 2001, p. 3.

model does not adequately reflect economic reality for knowledge-intensive businesses."

2. "This is, however, not easily remedied, since accounting adequately for knowledge-based business will ultimately require the invention of a new accounting model."[9]

The American Institute of Certified Public Accountants noted:

> Increased competition and rapid advances in technology are resulting in dramatic changes. To survive and compete, companies are changing everything—the way they are organized and managed, the way they do work and develop new products, the way they manage risks, and their relationships with other organizations . . . [They] are changing their information systems and the types of information they use to manage their businesses . . . Can business reporting be immune from the fundamental changes affecting business?[10]

Surprisingly, the AICPA recommended deferring any consideration of issues such as "accounting for intangible assets, including goodwill," "accounting for business combinations," and "alternative accounting principles."

At the "don't change anything" end of the spectrum, a 1997 magazine article expressed this view:

> The most troubling idea of the IC [intellectual capital] generation is to tinker with financial statements, so companies full of smart people who don't make profits look more attractive to investors. Some want to include the capitalized value of workers' ideas on the balance sheet. Some want to include cultural factors, such as the gender composition of the workforce, as if it is somehow a driver of the profitability of a company . . . Monkeying with financial statements, for almost any reason, is a terrible idea. Investors have 500 years of practice interpreting financial statements . . . [T]hey have developed methods to adjust for many of the anomalies (for example, amortiza-

[9] Robert I.G. McLean, *Performance Measures in the New Economy* (Canadian Institute of Chartered Accountants, Toronto, 1995). As reported in Financial Accounting Series No. 219-A, *Special Report: Business and Financial Reporting Challenges from the New Economy*, Wayne S. Upton, Jr. (Financial Accounting Standards Board, Apr. 2001), p. 13.

[10] *Improving Business Reporting—A New Customer Focus)* (AICPA, New York, 1994). As reported in Financial Accounting Series No. 219-A, *Special Report: Business and Financial Reporting Challenges from the New Economy*, Wayne S. Upton, Jr. Financial Accounting Standards Board, Apr. 2001), p. 10.

> tion of goodwill, which can only be defined by describing what it is not) that emerge from our archaic double-entry bookkeeping practices from time to time.[11]

FINANCIAL ACCOUNTING STANDARDS BOARD

Exposure Draft

Against this confusing backdrop, the FASB began the task of redesigning some critical accounting rules. On September 7, 1999, the FASB issued an Exposure Draft, "Proposed Statement of Financial Accounting Standards—Business Combinations and Intangible Assets." Written comments on this document were received until December 7, 1999, and public hearings were held in three cities in February 2000. The Exposure Draft was generally divided into two parts, the first concerning methods for accounting for business combinations, and the second concerning accounting for goodwill. The second portion is of more interest here and the following discussion focuses on that. The FASB highlighted several issues on which it was seeking guidance.

> *Issue 6*
> This proposed Statement would require that the excess of the cost of the acquisition price over the fair value of acquired net assets (goodwill) be recognized as an asset. This proposed Statement would require that goodwill be amortized over its useful economic life; however, the amortization may not exceed 20 years.[12]

On this subject, the FASB posed the following questions:

- Does goodwill meet the assets definition and the criteria for recognition as an asset in other FASB statements?
- Should goodwill be amortized in a manner similar to other assets?
- Is the 20-year maximum amortization period appropriate?

[11] John Rutledge, "You're a Fool If You Buy into This," *Forbes ASAP* (Apr. 1997). As reported in Financial Accounting Series No. 219-A, *Special Report: Business and Financial Reporting Challenges from the New Economy*, Wayne S. Upton, Jr. (Financial Accounting Standards Board, Apr. 2001), p. 4.

[12] This and the following statements of "Issues" are quoted from Financial Accounting Series No. 201-A, *Exposure Draft—Proposed Statement of Financial Accounting Standards* (Financial Accounting Standards Board, Sept. 7, 1999).

Issue 7
The Board considered several approaches that would have permitted some or all goodwill to be capitalized and not amortized. However, the Board found that none of these approaches were operational because of the subjectivity involved in identifying and measuring the discernible elements of goodwill, particularly those with indefinite lives and the inability to adequately review goodwill for impairment.

Relative to this issue, the FASB posed several questions:

- Can the subjectivity involved in identifying and measuring elements of goodwill be overcome?
- Is there a "robust and operational way" to review goodwill for impairment so that nonamortization of goodwill would be practical?

Issue 8
This proposed Statement would require acquired identifiable intangible assets that can be reliably measured to be recorded separately from goodwill in the financial statements of the acquiring enterprise at their fair value. That requirement is based on the assumption that intangible assets acquired in a business combination can be measured separately from goodwill with a sufficient degree of reliability to meet the asset recognition criteria. Based on information provided by valuation experts, the Board reached a conclusion that various intangible assets can be reliably measured.

The FASB asked for comment on several questions relating to this issue:

- Is the conclusion that intangible assets can be identified separate and apart from goodwill appropriate?
- Are some types of intangible assets more reliably measurable than others?
- Can the language in the proposed Statement be improved relative to recognizing intangible assets separate from goodwill?
- Are the examples of intangible assets shown in Appendix A of the Exposure Draft appropriate?

Issue 9
Opinion 17 imposed a 40-year maximum amortization period for all intangible assets. If certain criteria are met, this proposed Statement

would require an intangible asset (other than goodwill) to be amortized over a period longer than 20 years and in some circumstances to not be amortized at all.

Comments were sought relative to the following questions:

- Are the proposed criteria for extending the useful life beyond 20 years appropriate?
- Are the criteria for nonamortization appropriate?
- Are the illustrations given for the amortization period of certain identifiable intangible assets helpful?

The remaining issues in the Exposure Draft concerned the review of goodwill value for impairment, the method for reflecting goodwill amortization and impairment losses on the income statement, the disclosure of fair values assigned to intangible assets in notes to the financial statement, and the effective date and transition policy for application of the proposed Statement.

The Exposure Draft is a large document, two-thirds of which comprises an appendix presenting background information, the basis for the conclusions, and alternative views. This exemplifies the degree of controversy that existed about the issues, some of which were described earlier in this chapter. As to some of the issues discussed previously in Chapter 4, it is obvious that the FASB clearly intended to continue the exclusion of self-created intangible assets and intellectual property from the financial statements:

> Costs of internally developing, maintaining, or restoring intangible assets that are not specifically identifiable, have indeterminate lives, or are inherent in a continuing business and related to an enterprise as a whole shall be recognized as an expense when incurred.[13]

It is equally obvious that the FASB was struggling with the definition of goodwill and the question of whether to require amortization of the fair value assigned to it. Even though the proposed Statement calls for amortization of goodwill, the questions being asked indicate to us that the FASB was not fully decided on this point. It is not unusual for the FASB to find itself at odds with different groups of its constituency, depending upon the issue. This goodwill issue attracted more than normal controversy, however.

[13] This paragraph 36 of the Exposure Draft was carried forward from Opinion 17 without any change by the FASB.

It is clear that the FASB had a strong desire to eliminate the pooling-of-interest method of accounting for business combinations. An unusual circumstance caused this goal to become linked with the question of accounting for goodwill. When the acquisition of a business is accounted for as an asset purchase (nonpooling), the allocation of the purchase price usually identifies and assigns some value to goodwill. The amortization of this goodwill reduces the postacquisition earnings of the acquirer. This is an unpleasant prospect for the managers of that business. The managers of high-tech businesses were especially sensitive to this issue, because their acquisition targets were often other high-tech firms. Such businesses tend to have high enterprise value and low tangible asset value, indicating a large amount of intangible asset value (including goodwill). Business people in the high-tech segments argued that without pooling, the acquisition of such companies would be seriously impeded. The resulting large amount of goodwill, to be amortized over a maximum of 20 years as the FASB was proposing, would create an untenable burden of amortization for the acquiring company. In fact, some representatives of the high-tech business sector even importuned Congress to stop the FASB from doing away with pooling.

The FASB was therefore in somewhat of a dilemma, knowing that a resolution of this issue was key to obtaining its objective of eliminating pooling. Thus, it raised these questions concerning the identification and valuation of goodwill, its amortization, and the possibility of using impairment testing as a substitute for amortization.

It also appears to us that the FASB was seeking to achieve a more definitive identification and valuation of intangible assets and intellectual property resulting from a business acquisition. Opinions 16 and 17 (described more fully in Chapter 4) have always required an identification of intangible assets. In our experience, however, this identification was often not carried out in a rigorous fashion. The extent to which values were assigned to acquired intangible assets and intellectual property was often influenced by the desire of management of the acquiring company to assign as much value as possible to goodwill, with its then longer period of amortization. Or, in the case in which the amortization period for some intangibles was about the same as for goodwill, there simply was no reason to identify them separately. Thus, allocations of purchase price often resulted in a one-line balance sheet entry: "Goodwill and Other Intangible Assets." In the Exposure Draft, the FASB was clearly desirous of more detailed information, even though the suggestion was that it would be contained in the notes to financial statements.

Perhaps another reason for the FASB's focus on a rigorous identification of intangibles was its realization that nonamortization of goodwill was a distinct possibility. If that happened, there would be a strong motivation for business people to allocate as much as possible to this "free" goodwill asset.

The Exposure Draft defined *intangible assets* as "noncurrent assets (not including financial instruments) that lack physical substance." *Goodwill* was defined as consisting "of one or more unidentified intangible assets and identifiable intangible assets that are not reliably measurable."[14] In an appendix to the Exposure Draft, some guidance was provided by a suggested list of intangible assets. We include that list as Appendix 4A to this chapter. The FASB then suggested criteria for separately recognizing an intangible asset:

- Is the intangible asset identifiable?
- Can its value be reliably measured?

The Exposure Draft then suggested that although intangibles should be amortized over their useful economic lives, that life is presumed to be 20 years or less. However, amortization can be made over a longer period if certain conditions are met:

- The intangible asset must be exchangeable. Examples include a workforce or customer relationships which can only generate cash flows by using them, rather than by disposing of them. (¶¶ 284 and 285).
- If the intangible asset is not exchangeable, it must be controlled by contractual or legal rights that extend beyond 20 years.
- The intangible asset, or a group of intangibles that includes it, must have a clearly identifiable cash flow stream extending beyond 20 years.
- The intangible asset must have a finite useful economic life.

If the intangible asset passes the first three of these tests, but cannot be shown to have a finite useful economic life, then it may qualify as *not* amortizable:

- To be nonamortizable, the intangible asset must have an "observable market." An *observable market* is "one in which intangible assets are separately bought and sold, even though such transactions may be infrequent. From those purchase and sale transactions, a market price can be observed and used in estimating the fair value of intangible assets that are similar" (¶34).

[14] Exposure Draft, paragraph 34.

The FASB was clearly opening the door to consideration of recognizing intangible assets that need not be amortized, but it kept the 20-year chain on the lock for goodwill. We will observe how these issues were resolved in the final Statements.

Statements of Financial Accounting Standards Nos. 141 and 142

The matters addressed in the Exposure Draft were finalized in Statements of Financial Accounting Standards No. 141 (Business Combinations) and No. 142 (Goodwill and Other Intangible Assets). The following discussion relates to both of these documents, as some of the same subject matter is addressed in both. Notations are included to identify from which Standard the quotations are taken, and the cited paragraph numbers are those of the documents.

Purchase Price Allocation

At the most general level, we are told that the pooling-of-interest method of accounting for a business combination is no longer permitted and that the purchase consideration for a group of assets must be allocated to those assets based on their fair value. This is to include intangible assets and goodwill. We are further instructed that goodwill is to be valued on a residual basis and only in the case of a business combination:

> 7. *Allocating cost.* Acquiring assets in groups requires not only ascertaining the cost of the asset . . . group but also allocating that cost to the individual assets . . . that make up the group. The cost of such a group is determined using the concepts described in paragraphs 5 and 6.[15] A portion of the cost of the group is then assigned to each individual asset . . . acquired on the basis of its fair value. In a business combination, an excess of the cost of the group over the sum of the amounts assigned to the tangible assets, financial assets, and separately recognized intangible assets acquired less liabilities assumed is evidence of an unidentified intangible asset or assets.[16] (SFAS 141)

[15] The fair values of the net assets acquired and the consideration paid are assumed to be equal. The "cost" of an asset in this case is assumed to be equal to its market value.

[16] This and other quotations identified by a paragraph number are from SFAS 141 or SFAS 142. We have added occasional emphasis by italicizing text.

> 9. An intangible asset that is acquired either individually or with a group of other assets (but not those acquired in a business combination) shall be initially recognized and measured based on its fair value . . . The cost of a group of assets acquired in a transaction other than a business combination shall be allocated to the individual assets acquired based on their relative fair values and shall not give rise to goodwill. (SFAS 142)

Value Premise

In Table 4A.1 we show the assets and liabilities typically acquired in a business combination, together with the valuation premise for each. This table replaces Table 4.1 that appears in Chapter 4 of the main volume.

Defining Intangible Assets

In our discussion of the Exposure Draft, we commented on the FASB's struggle to define *identifiable intangible assets* and to differentiate them from goodwill. SFAS 141 considerably refines the criteria for recognizing an intangible asset separate and apart from goodwill. The primary criterion is that the asset arise from contractual or other legal rights. Examples given include a favorable lease contract, the license to operate a nuclear power plant, and a patent.

If an acquired intangible asset does *not* arise from contractual or legal rights, it can be identified as a separate asset if it is capable of being separated or divided from the acquired entity and sold, transferred, licensed, rented, or exchanged. This criterion can be met irrespective of whether the acquiring entity intends to enter into any of these transactions. Examples given are customer and subscriber lists that are frequently leased. Further, an asset can meet the separability criterion even if it might be separable only in combination with other assets, such as a trademark, related secret formula, or proprietary technology used to manufacture a single product. All three of these assets would meet the separability criterion because they belong to a separable "package."

> 39. An intangible asset shall be recognized as an asset apart from goodwill if it *arises from contractual or other legal rights* (regardless of whether those rights are transferable or separable from the acquired entity or from other rights and obligations). If an intangible asset does not arise from contractual or other legal rights, it shall be recognized as an asset apart from goodwill *only if it is separable*,

Table 4A.1 Value Premises for Allocation of Purchase Consideration

Item	Value Premise
Assets	
Current Assets	
Marketable securities	FV
Accounts receivable	PV
Inventories	
Finished goods	SP-(D+P)
Work in progress	SP-(C+D+P)
Raw materials	RC
Plant and Equipment	
To be used	RCNLD
To be sold	FV-D
Intangible Assets	FV
Other Assets	FV
Liabilities	
Accounts and Notes Payable	PV
Long-Term Debt	PV
Liabilities Associated with Pension or Post-Retirement Plans	see FASB Statements 87 & 106
Accruals	PV
Other Liabilities and Commitments	PV

Where:

PV =	Present value at a current interest rate reflective of risk of receiving the income, less allowances, if appropriate
FV =	Fair value
SP =	Selling price
C =	Completion cost
P =	Profit
D =	Disposal cost
RC =	Current replacement cost
RCNLD =	Current replacement cost less depreciation or used-asset market value

that is, it is capable of being separated or divided from the acquired entity and sold, transferred, licensed, rented, or exchanged (regardless of whether there is an intent to do so). For purposes of this Statement, however, an intangible asset that cannot be sold, transferred, licensed, rented, or exchanged individually *is considered separable if it can be sold, transferred, licensed, rented, or exchanged in combination with a related contract, asset, or liability*. For purposes of this Statement, *an assembled workforce shall not be recog-*

> *nized as an intangible asset apart from goodwill.* Appendix A [to SFAS 141] provides additional guidance relating to the recognition of acquired intangible assets apart from goodwill, including an illustrative list of intangible assets that meet the recognition criteria in this paragraph. (SFAS 141)

This separability standard is one that the reader will recognize as being absent from our presentations in this book. We (and a number of other valuation professionals who commented on the Exposure Draft) do not regard separability as a defining characteristic of an intangible asset. We cited an assembled workforce as an example. Our comments were in vain, however. When these Statements were issued we learned why. The FASB did not consider replacement cost a reliable measure of value and, because that is the valuation method of choice for an assembled workforce, the Board directed that the asset be considered part of goodwill:

> 43. The excess of the cost of an acquired entity over the net of the amounts assigned to assets acquired and liabilities assumed shall be recognized as an asset referred to as goodwill. An acquired intangible asset that does not meet the criteria in paragraph 39 shall be included in the amount recognized as goodwill. (SFAS 141)

SFAS 141 (paragraph a14) provides a new list of intangible assets that would be expected to meet the criteria for recognition apart from goodwill. We have reproduced this list as Appendix 4B to this chapter. This new list is organized into five asset groups: *marketing-related, customer-related, artistic-related, contract-based*, and *technology-based*. This better organization arose, we believe, from the comments received by the FASB relative to the list presented as Appendix 4A of the Exposure Draft. Two of the asset categories contained in the original list—workforce-based assets and corporate organizational and financial assets—do not appear in the final list of examples. For the most part, the assets listed under these classifications would not meet the new criteria and would therefore be lumped with goodwill.

As expected, the FASB confirmed its decision to exclude any reflection of self-created intangible assets or intellectual property from the financial statements:

> 10. Costs of internally developing, maintaining, or restoring intangible assets (including goodwill) that are not specifically identifiable, that have indeterminate lives, or that are inherent in a continuing business and related to an entity as a whole, shall be recognized as an expense when incurred. (SFAS 142)

Useful Economic Life

On the subject of economic life, SFAS 142 essentially retained the concepts introduced in the Exposure Draft. The reader can compare the economic life analysis described in the excerpt quoted here with the concepts presented in Chapter 10 in the main volume:

> 11. The accounting for a recognized intangible asset is based on its useful life to the reporting entity. An intangible asset with a finite useful life is amortized; an intangible asset with an indefinite useful life is not amortized. The *useful life of an intangible asset to an entity is the period over which the asset is expected to contribute directly or indirectly to the future cash flows of that entity.* The estimate of the useful life of an intangible asset to an entity shall be based on an analysis of all pertinent factors, in particular:
>
> a. The expected use of the asset by the entity
> b. The expected useful life of another asset or a group of assets to which the useful life of the intangible asset may relate (such as mineral rights to depleting assets)
> c. Any legal, regulatory, or contractual provisions that may limit the useful life
> d. Any legal, regulatory, or contractual provisions that enable renewal or extension of the asset's legal or contractual life without substantial cost (provided there is evidence to support renewal or extension and renewal or extension can be accomplished without material modifications of the existing terms and conditions)
> e. The effects of obsolescence, demand, competition, and other economic factors (such as the stability of the industry, known technological advances, legislative action that results in an uncertain or changing regulatory environment, and expected changes in distribution channels)
> f. The level of maintenance expenditures required to obtain the expected future cash flows from the asset (for example, a material level of required maintenance in relation to the carrying amount of the asset may suggest a very limited useful life).[17]
>
> If no legal, regulatory, contractual, competitive, economic, or other factors limit the useful life of an intangible asset to the reporting en-

[17] As in determining the useful life of depreciable tangible assets, regular maintenance may be assumed but enhancements may not.

tity, the useful life of the asset shall be considered to be indefinite. The term *indefinite* does not mean infinite. (SFAS 142)

Most of these suggestions are useful, with the possible exception of the last item. It does not seem to us that the level of maintenance expenditures attributed to an asset is relevant to a judgment about its economic life. Such maintenance expenditures are made for the purpose of prolonging life and, once made, prolong life.

Item d. is noteworthy. The FASB has correctly recognized that some intangible assets, such as licenses, franchises, and certifications, may have an economic life *longer* than their legal life. Many such assets are routinely renewed at little or no cost, as long as compliance remains.

Also noteworthy is the subject of economic life as it relates to the category of marketing-related intangibles, which includes:

- Trademarks, trade names
- Service marks, collective marks, certification marks
- Trade dress (unique color, shape, or package design)
- Newspaper mastheads
- Internet domain names
- Noncompetition agreements

Of the assets in this group, only noncompetition agreements would commonly have a finite life, defined by contract. The other assets would very likely be judged to have indefinite lives and thus not be amortized. We must be careful not to adopt this as the "conventional wisdom,"[18] however, but it will often be true. With these recent Statements there will, for the first time, be a strong motivation to identify and properly value these trademark assets in an acquisition.

More importantly, SFAS 142 echoes the concept introduced in the Exposure Draft in its specification that "the method of amortization shall reflect the *pattern* in which the economic benefits of the intangible assets are consumed or otherwise used up" (¶12). As we noted in the main volume of this book (Chapter 10 and Appendix B), most intangible assets in fact do deteriorate in value over a pattern that is not a straight line. This fact was given a great deal of attention in

[18] Although their legal life is assumed to be perpetual, trademarks often stop producing cash flow for their owners. They are subject to economic, functional, event, technological, product, and cultural obsolescence. See Gordon V. Smith, *Trademark Valuation* (New York: John Wiley & Sons, 1997), ch. 5.

the past relative to the amortization of intangible assets for tax purposes. There were many confrontations between taxpayers and the Internal Revenue Service and some were extensively litigated. There is a considerable body of knowledge extant on this subject. The need for that attention was legislated out of existence in the tax law, but is now being revived in these new accounting requirements.

The SEC has offered its own comment relative to the accounting for customer relationship intangibles, which were often the focal point of the taxpayer / IRS disputes noted here:

> Some intangible assets recognized in a purchase business combination derive their value from future cash flows expected to be derived from the acquired business' identified customers. Companies may also recognize this type of intangible asset when they acquire groups of customer accounts or a customer list. Most commonly, valuable continuing relationships are demonstrated by existing contracts or subscriptions.
>
> When acquired in a business combination or as part of a larger group of assets, the fair value of this intangible is often measured as the present value of the estimated net cash flows from the contracts, including expected renewals. The most reliable indication of life expectancy of a subscriber base or similar customer group is the historical life experience of similar customer accounts. The actuarial-based retirement rate is the method generally accepted in the appraisal profession to estimate life expectancy. That analysis may be developed if customer initiation and termination data are maintained for each acquired customer group.
>
> Typically, customer relationships within a large group of accounts tend to dissipate at a more rapid rate in the earlier periods following a company's succession to the contracts, with the rate of attrition declining over time until relatively few customers remain who persist for an extended period. Under this pattern, the preponderance of cash flows derived from the acquired customer base will be recognized in income in the earlier periods, and they fall to a materially reduced level in later years. In this circumstance, straight-line cost amortization over the period of expected cash flows particularly will exaggerate net earnings when the business is growing, leaving disproportionate expense to be recognized when the rate of growth declines. The staff believes that an accelerated method of amortization, rather than the straight-line method, will result in the most appropriate and systematic allocation of the intangible's cost to the periods benefited. The straight-line method is appropriate only if the estimated life of the intangible assets is shortened to assure that recognition of the cost of

> the revenues, represented by amortization of the intangible asset, better corresponds with the distribution of expected revenues."[19]

Goodwill Amortization

We can now observe how the FASB resolved the dilemma that we highlighted in our discussion of the Exposure Draft. The FASB eliminated the amortization of goodwill and in doing so eased the pain of acquiring companies caused by loss of the pooling methodology. Understandably, the FASB could not accept the idea that goodwill, once recorded, would reside on the balance sheet of the acquiring company forever. Thus, the FASB extended the concept of measuring impairment to the goodwill asset.

Measuring Impairment

The measurement of goodwill impairment is somewhat complex and involves the establishment of so-called "reporting units." A reporting unit within a company is an organization with characteristics similar to a business segment and for which separate financial information is available and for which there is a management team that reviews the operating results. Goodwill, including goodwill that exists in the financial statements as of the effective date of these statements, must be allocated to reporting units.

The goodwill assigned to a reporting unit must be tested for impairment at least annually and more often if there is an event that would affect the reporting unit. Such events might include an adverse business climate or litigation, loss of key personnel, unanticipated competition, or other events that could be expected to be detrimental to the business health of the reporting unit. Impairment testing proceeds in a two-step process:

> 18. Goodwill shall not be amortized. Goodwill shall be tested for impairment at a level of reporting referred to as a reporting unit . . . Impairment is the condition that exists when the carrying amount of goodwill exceeds its implied fair value.[20] The *two-step impairment*

[19] SEC Division of Corporation Finance, "Current Accounting and Disclosure Issues," prepared by a member of the staff, Aug. 31, 2001. See <www.sec.gov/divisions/corpfin/acctdisc.htm>.

[20] The fair value of goodwill can be measured only as a residual and cannot be measured directly. Therefore, this Statement includes a methodology to determine an amount that achieves a reasonable estimate of the value of goodwill for purposes of measuring an impairment loss. That estimate is referred to herein as the *implied fair value of goodwill*.

> *test* discussed in paragraphs 19-22 shall be used to identify potential goodwill impairment and measure the amount of a goodwill impairment loss to be recognized (if any). (SFAS 142)

Step One

> 19. The first step of the goodwill impairment test, used to identify potential impairment, *compares the fair value of a reporting unit with its carrying amount, including goodwill.* The guidance in paragraphs 23-25 shall be used to determine the fair value of a reporting unit. If the fair value of a reporting unit exceeds its carrying amount, goodwill of the reporting unit is considered not impaired, thus the second step of the impairment test is unnecessary. If the carrying amount of a reporting unit exceeds its fair value, the second step of the goodwill impairment test shall be performed to measure the amount of impairment loss, if any. (SFAS 142)

Step Two

> 20. The second step of the goodwill impairment test, used to measure the amount of impairment loss, *compares the implied fair value of reporting unit goodwill with the carrying amount of that goodwill.* [See ¶21.] If the carrying amount of reporting unit goodwill exceeds the implied fair value of that goodwill, an impairment loss shall be recognized in an amount equal to that excess . . . After a goodwill impairment loss is recognized, the adjusted carrying amount of goodwill shall be its new accounting basis. (SFAS 142)
>
> 21. The implied fair value of goodwill shall be determined in the same manner as the amount of goodwill recognized in a business combination is determined. That is, an entity shall allocate the fair value of a reporting unit to all of the assets and liabilities of that unit (including any unrecognized intangible assets) as if the reporting unit had been acquired in a business combination and the fair value of the reporting unit was the price paid to acquire the reporting unit. The excess of the fair value of a reporting unit over the amounts assigned to its assets and liabilities is the implied fair value of goodwill. (SFAS 142)

The starting point in the impairment test is to establish the fair value of each reporting unit. If that fair value substantially exceeds the book value (*carrying value*) of all the assets, including goodwill, it is assumed that there is no impairment and the process may stop there.

If the sum of book values is close to, or higher than, the fair value of the reporting unit, then some impairment is assumed. It becomes necessary to appraise all of the tangible and identifiable intangible assets. The sum of these values is subtracted from the unit value to calculate the value of "implied goodwill." If the implied goodwill is less than the value of goodwill on the books of the unit, the difference is the impairment loss. The implied goodwill value becomes the new book value of goodwill in the unit. The other assets are not revalued.

Because there is some latitude in the selection of what will be "reporting units," most managers will make the selection strategically.[21] Companies with existing goodwill must allocate it to reporting units. In doing so, however, they may discover an initial goodwill impairment.[22] Such an impairment loss may be reflected in the income statement as a loss due to a "change of accounting." This is more palatable than a future impairment discovery.

A goodwill impairment loss in the future must go through the income statement as an operating loss—not an attractive prospect. Managers will be attentive to how much of the existing goodwill the contemplated reporting units will attract in the allocation process. They will be evaluating the potential value of identifiable intangibles a reporting unit is likely to have. They will analyze this in the light of the business outlook for a unit.

SFAS 141 and 142 apply initially only to companies that have goodwill on their books. As companies make acquisitions, they will come under these rules.

Disclosure Requirements

One of the more interesting portions of SFAS 141 and 142 is the disclosure requirements. If the companies that are subject to these requirements closely follow the disclosure specifications, much useful information will become available following their acquisitions. Obviously the intent of the FASB was to cause this information to become available to the companies' stakeholders–investors and lenders. This is in accordance with the original impetus for these new requirements. As valuation professionals, we are interested in the availability of this information in that it will provide additional data points relating to how other professionals have valued intangible assets and intellectual property.

> 44. For intangible assets acquired either individually or with a group of assets, the following information shall be disclosed in the notes to the financial statements in the period of acquisition:

[21] SFAS 142, ¶35.
[22] SFAS 142, ¶56.

a. For intangible assets *subject to amortization*:
   (1) The total amount assigned and the *amount assigned to any major intangible asset class*
   (2) The amount of any significant *residual value*, in total and by major intangible asset class
   (3) The *weighted-average amortization period*, in total and by major intangible asset class
b. For intangible assets *not subject to amortization*, the total amount assigned and the amount assigned to any major intangible asset class
c. The *amount of research and development* assets acquired and written off in the period and the line item in the income statement in which the amounts written off are aggregated.

45. The following information shall be disclosed in the financial statements or the notes to the financial statements for each period for which a statement of financial position is presented:

a. For intangible assets subject to amortization:
   (1) The gross carrying amount and accumulated amortization, in total and by major intangible asset class
   (2) The aggregate amortization expense for the period
   (3) The estimated aggregate amortization expense for each of the five succeeding fiscal years
b. For intangible assets *not* subject to amortization, the total carrying amount and the carrying amount for each major intangible asset class
c. The changes in the carrying amount of goodwill during the period including:
   (1) The aggregate amount of goodwill acquired
   (2) The aggregate amount of impairment losses recognized
   (3) The amount of goodwill included in the gain or loss on disposal of all or a portion of a reporting unit.

Entities that report segment information in accordance with Statement 131 shall provide the above information about *goodwill in total and for each reportable segment* and shall disclose any significant changes in the allocation of goodwill by reportable segment. If any portion of goodwill has not yet been allocated to a reporting unit at the date the financial statements are issued, that unallocated amount and the reasons for not allocating that amount shall be disclosed. (SFAS 142)

SEC Accounting Staff members have made additional suggestions relative to disclosures about intangible assets. Some of these are rather extreme, but indicate the direction of their thinking:

Registrants should consider the need for more extensive narrative and quantitative information about the intangibles that are important to their business. These disclosures often are appropriate in *Description of Business* or *Management's Discussion & Analysis*. Some disclosures required by GAAP or Commission rules provide useful information to investors about intangibles, such as amounts annually expended for advertising and research & development. More insight could be provided if management elected to disaggregate those disclosed amounts by project or purpose. Statistics about workforce composition and turnover could highlight the condition of that human resource intangible. Disclosure of annual expenditures relating to training and new technologies could help investors distinguish one company's intangibles from another. More specific information about patents, copyrights, and licenses, including their duration, royalties, and competitive risks can be important to investors. Insight into the intangible value of management talent could be provided by supplementing financial information with performance measures used to assess management's effectiveness.[23]

FASB Proposed Project

The FASB, in August 2001, issued a Request for Comments on a proposal for a project on disclosure about intangibles. Comments were due by October 5, 2001. This document recognizes that "intangible assets are generally recognized only if acquired, either separately or as part of a business combination. Intangible assets that are generated internally, and some acquired assets that are written off immediately after being acquired, are not reflected in financial statements, and little quantitative or qualitative information about them is reported in the notes to the financial statements."[24] The FASB describes this proposed project as having two goals:

1. Make new information available to investors and creditors and improve the quality of information currently being provided.
2. Take the first step in what might become an evolution toward recognition in an entity's financial statements of internally generated intangible assets.

[23] SEC Division of Corporation Finance, "Current Accounting and Disclosure Issues," prepared by a member of the staff, Aug. 31, 2001. See <www.sec.gov/divisions/corpfin/acctdisc.htm>.
[24] Proposal for a New Agenda Project, "Disclosure of Information About Intangible Assets Not Recognized in Financial Statements," Financial Accounting Standards Board, p. 1.

One of the interesting comments made by the FASB in this proposal is that current accounting practices make it "difficult to compare the financial statements of an entity that has built up substantial intangible assets internally with those of another entity that has purchased most of its intangible assets." The FASB notes that investors and lenders could make more meaningful comparisons between companies if heretofore unrecognized intangibles were disclosed in the financial statements.

The proposed project would focus on four issues:

1. What intangible assets are to be included?
   The proposed scope includes intangible assets that are not currently recognized, but that would have been recognized had they been acquired from others. Also to be considered would be in-process research and development assets written off immediately after an acquisition.
2. What information should be disclosed about intangible assets?
   In this regard, the project proposes potential quantitative and qualitative disclosures:
   - Major classes of intangible assets and their characteristics;
   - Expenditures to develop and maintain them;
   - Value of those assets;
   - Significant events that could change the anticipated future benefits arising from intangible assets.
3. Should the disclosures be voluntary or required?
   The FASB recognized that some industry groups might be more likely than others to volunteer disclosures. That might reduce the resistance to this type of disclosure. The FASB also recognized that voluntary participation might be very limited.
4. Should the disclosures be made annually or more frequently?

By limiting the focus of this project to intangible assets that are not recognized currently, but would be recognized if acquired separately or in a business combination, the FASB has made this project more feasible. The FASB also pointed out, however, that it had considered, and rejected, additional scope for the project that included:

1. "Disclosure of non-financial indicators about intangible factors, such as market size and share, customer satisfaction levels, new product success rates, and employee retention rates."

2. "Recognition and measurement, in statements of financial position, of research and development and other project-related intangible assets."
3. "Separate recognition and measurement of intangible assets or liabilities embedded in tangible or financial assets, for example, banks' core deposit intangibles and insurers' claim-handling obligations."

We can envision considerable resistance on the part of companies to disclose some of this information, and it is well that the FASB rejected going into these areas at this time. There will be resistance enough to what has been proposed. It will be most interesting to observe the responses to this proposed project. Stay tuned.

Appendix 4A

Intangible Asset List—Exposure Draft

Customer-based or market-based assets—intangible assets that relate to customer structure or market factors of the business:

a. Lists (advertising, customer, dealer, mailing, subscription, and so forth)

b. Customer base

c. Financial institution depositor or borrower relationships

d. Customer routes

e. Delivery system, distribution channels

f. Customer service capability, product or service support

g. Effective advertising programs

h. Trademarked brand names

i. Newspaper mastheads

j. Presence in geographic locations or markets

k. Value of insurance-in-force, insurance expirations

l. Production backlog

m. Concession stands

n. Airport gates and slots

o. Retail shelf space

p. Files and records (credit, medical)

Contract-based assets—intangible assets that have a fixed or definite term:

a. Agreements (consulting, income, licensing, manufacturing, royalty, standstill)

b. Contracts (advertising, construction, consulting, customer, employment, insurance, maintenance, management, marketing, mortgage, presold, purchase, service, supply)

c. Covenants (not to compete)

d. Easements

e. Leases (valuable or favorable terms)

f. Permits (construction)

g. Rights (broadcasting, development, gas allocation, landing, lease, mineral, mortgage servicing, reacquired franchise, servicing, timber cutting, use, water)

Technology-based assets—intangible assets that relate to innovations or technological advances within the business:

a. Computer software and license, computer programs, information systems, program formats, Internet domain names and portals

b. Secret formulas, processes, and recipes

c. Technical drawings, technical and procedural manuals, blueprints

d. Databases, title plants

e. Manufacturing processes, procedures, production line

f. Research and development

g. Technological know-how

Statutory-based assets—intangible assets with statutorily established useful lives:

a. Patents

b. Copyrights (manuscripts, literary works, musical compositions)

c. Franchises (cable, radio, television)

d. Trademarks, trade names

Workforce-based assets—intangible assets that relate to the value of the established employees or workforce of a company:

a. Assembled workforce, trained staff

b. Nonunion status, strong labor relations, favorable wage rates

c. Superior management or other key employees

d. Technical expertise

e. Ongoing training programs, recruiting programs

Corporate organizational and financial assets—intangible assets relating to the organizational structure of an entity:

a. Savings value of escrow fund

b. Favorable financial arrangements, outstanding credit rating

c. Fundraising capabilities, access to capital markets

d. Favorable government relations

Appendix 4B

Intangible Asset List—Final SFAS No. 141

a. Marketing-related intangible assets

(1) Trademarks, trade names

(2) Service marks, collective marks, certification marks

(3) Trade dress (unique color, shape, or package design)

(4) Newspaper mastheads

(5) Internet domain names

(6) Noncompetition agreements

b. Customer-related intangible assets

(1) Customer lists ▲

(2) Order or production backlog

(3) Customer contracts and related customer relationships

(4) Noncontractual customer relationships ▲

▲ Denotes assets that do not arise from contractual or other legal rights, but are recognized because they meet the separability criterion.

All other assets meet the contractual/legal criterion.

c. Artistic-related intangible assets

(1) Plays, operas, ballets

(2) Books, magazines, newspapers, other literary works

(3) Musical works such as compositions, song lyrics, advertising jingles

(4) Pictures, photographs

(5) Video and audiovisual material, including motion pictures, music videos, television programs

d. Contract-based intangible assets

(1) Licensing, royalty, standstill agreements

(2) Advertising, construction, management, service, or supply contracts

(3) Lease agreements

(4) Construction permits

(5) Franchise agreements

(6) Operating and broadcast rights

(7) Use rights such as drilling, water, air, mineral, timber cutting, and route authorities

(8) Servicing contracts such as mortgage servicing contracts

(9) Employment contracts

e. Technology-based intangible assets

(1) Patented technology

(2) Computer software and mask works

(3) Unpatented technology ▲

(4) Databases, including title plants ▲

(5) Trade secrets, such as secret formulas, processes, recipes

▲ Denotes assets that do not arise from contractual or other legal rights, but are recognized because they meet the separability criterion.

All other assets meet the contractual/legal criterion.

Appendix 4C

Relevant Documents in the Development of SFAS No. 141 and SFAS No. 142

No. 201-A: Financial Accounting Series—September 7, 1999
Exposure Draft: Proposed Statement of Financial Accounting Standards
Business Combinations and Intangible Assets

Comment deadline: December 7, 1999
Hearings: February 3 and 4, 2000, in San Francisco, CA
February 8, 2000, in Norwalk, CT
February 10 and 11, 2000, in New York, NY

Independence Standards Board (ISB)—September 1999
Discussion Memorandum DM 99-3
Appraisal and Valuation Services

No. 219-A: Financial Accounting Series—April, 2001
Special Report
Business and Financial Reporting, Challenges from the New Economy

Wayne S. Upton, Jr.

Financial Accounting Series—May 17, 2001
Statement of Financial Accounting Standards No. 142
Business Combinations

Confidential Draft
Later became No. 141

No. 221-B: Financial Accounting Series—June 2001
Statement of Financial Accounting Standards No. 141
Business Combinations

Published by Financial Accounting Standards Board of the Financial Accounting Foundation

Supersedes APB Opinion No. 16, *Business Combinations*, and FASB Statement No. 38, *Accounting for Preacquisition Contingencies of Purchased Enterprises.*

No. 221-C: Financial Accounting Series—June 2001
Statement of Financial Accounting Standards No. 142
Goodwill and Other Intangible Assets

Supersedes APB Opinion No. 17, *Intangible Assets.*

*11A (New)

New Challenges for the Expert Witness

Recent court decisions have put the spotlight on expert witness testimony. Most notable is the 1993 U.S. Supreme Court decision in the *Daubert*[1] case, which tightened considerably the standards for "scientific" expert testimony in federal court. Just as appraisers, accountants, and economists were breathing a sigh of relief that the new standards did not apply to them, the Court's decision in *Kumho Tire*[2] extended the new standards to expert testimony concerning "technical" or "other specialized" knowledge. Though it is unlikely that readers in the legal and valuation professions are unaware of these cases, the implications of the decisions, as they specifically relate to valuation and damage issues, bear closer examination.

It is not our intent to debate the legal or social merits of *Daubert*[3] and the *Daubert*-like decisions that have been rendered, but rather to present a discussion of the issues so that readers can:

[1] *Daubert v. Merrell Dow Pharmaceuticals, Inc.*, 509 US 579, 125 L Ed 2d 469, 113 S Ct 2786 (1993).
[2] *Kumho Tire Co. v. Carmichael*, 526 US 137, 119 S Ct 1167, 1999 US LEXIS 2189 (Mar 23, 1999).
[3] *See, e.g.*, Sharon Begley, "Ban on 'Junk Science' Also Keeps Jurors from Sound Evidence," *Wall Street Journal*, June 27, 2003.

- Better understand the implications of these decisions for professional valuation activities, and
- Better understand how to avoid failing these tests in litigation involving intangible asset and intellectual property valuations and damage quantification.

It has been suggested that because *Daubert* was a federal case, and was adopted by only some states, its result would not affect litigation in other courts. It may also be that the *Daubert* principles will be expanded or supplanted by future decisions. We feel, however, that expert witnesses will continue to be required to meet ever-higher standards as science and expertise come before the bar, whatever the jurisdiction. We have, as an example, observed that standards promulgated for international transfer pricing in the tax arena have often been adopted by U.S. state courts in income tax matters.

THE CASES: *DAUBERT* AND *KUMHO*

What Is Daubert About?

The *Daubert* case is described as follows:

> Petitioners, two minor children and their parents, alleged in their suit against Respondent that the childrens' serious birth defects had been caused by the mother's prenatal ingestion of bendectin, a prescription drug marketed by Respondent. The District Court granted Respondent summary judgment based on a well-credentialed expert's affidavit concluding, upon reviewing the extensive published scientific literature on the subject, that maternal use of bendectin has not been shown to be a risk factor for human birth defects. Although petitioners had responded with the testimony of eight other well-credentialed experts, who based their conclusion that bendectin can cause birth defects on animal studies, chemical structure analyses, and the unpublished "reanalysis" of previously published human statistical studies, the Court determined that this evidence did not meet the applicable "general acceptance" standard for the admission of expert testimony.[4]

[4] *Daubert*, 509 US at Syllabus, p. 1.

The U.S. Court of Appeals for the Ninth Circuit affirmed the decision, citing *Frye*,[5] which for 70 years had been the primary guide for expert opinions; *Frye* states that an expert opinion based on a scientific technique is inadmissible unless the technique is "generally accepted." Questions of admissibility under this guidance were often resolved in a statement by the court to the effect that "the jury will consider the weight of the evidence"; in other words, "Go ahead and present your testimony (and be cross-examined) and we'll see to what extent it influences the jury's decision."

The Supreme Court noted that the *Frye* test had been the subject of much debate over the years and agreed with those debaters who asserted that *Frye* had been supplanted by the Federal Rules of Evidence, citing Rules 702 and 703:

> Rule 702. Testimony by Experts
> If scientific, technical, or other specialized knowledge will assist the trier of fact to understand the evidence or to determine a fact in issue, a witness qualified as an expert by knowledge, skill, experience, training, or education, may testify thereto in the form of an opinion or otherwise, if (1) the testimony is based upon sufficient facts or data, (2) the testimony is the product of reliable principles and methods, and (3) the witness has applied the principles and methods reliably to the facts of the case.
>
> Rule 703. Bases of Opinion Testimony by Experts
> The facts or data in the particular case upon which an expert bases an opinion or inference may be those perceived by or made known to the expert at or before the hearing. If of a type reasonably relied upon by experts in the particular field in forming opinions or inferences upon the subject, the facts or data need not be admissible in evidence in order for the opinion or inference to be admitted. Facts or data that are otherwise inadmissible shall not be disclosed to the jury by the proponent of the opinion or inference unless the court determines that their probative value in assisting the jury to evaluate the expert's opinion substantially outweighs their prejudicial effect.

The Court held that "the Federal Rules of Evidence, not *Frye*, provide the standard for admitting expert scientific testimony in a federal trial."[6] The Justices commented that the Rules of Evidence, in displacing Frye, did not imply that there are few limits on the admissibility of "purportedly scientific evidence," but

[5] *Frye v. United States*, 54 App DC 46, 47, 293 F 1013, 1014 (1923).
[6] *Daubert*, 509 US at Syllabus, p. 1.

rather that "the trial judge must ensure that any and all scientific testimony or evidence admitted is not only relevant, but reliable."[7]

To summarize, the current standards for expert testimony require that it:

- Be based upon sufficient facts or data,
- Be the product of reliable principles and methods, and
- Apply those principles and methods reliably to the facts of the case.

The *Daubert* Court continued with some general observations about tests that could be employed to enable a conclusion about whether a theory or technique is "scientific . . . knowledge [that] will assist the trier of fact"[8]:

- Can it be tested?
- Has it been subjected to peer review and publication?
- Can the rate of error be known?
- Is there widespread acceptance in the relevant community?

In its *Daubert* decision, the Supreme Court stated:

> To summarize: "General acceptance" is not a necessary precondition to the admissibility of scientific evidence under the Federal Rules of Evidence, but the Rules of Evidence—especially Rule 702—do assign to the trial judge the task of ensuring that an expert's testimony both rests on a reliable foundation and is relevant to the task at hand. Pertinent evidence based on scientifically valid principles will satisfy those demands.[9]

The judgment of the court of appeals was vacated and remanded, in that its decision (and that of the district court) was "focused almost exclusively on 'general acceptance' as gauged by publication."[10]

Then Came *Kumho*

On July 6, 1993, the *Kumho* plaintiff was driving a minivan on which a rear tire blew out. In the ensuing accident, one of the passengers died and others were se-

[7] *Id.* at p. 9.
[8] *Id.* at p. 9.
[9] *Id.* at p. 18.
[10] *Id.* at p. 18.

verely injured. Suit was brought against the tire maker and its distributor, claiming that the tire was defective. A significant element of the plaintiff's case was extensive deposition testimony provided by an expert in tire failure analysis.

The case was heard in district court. The defendant moved to exclude the expert's testimony on the ground that his methodology failed Rule 702's reliability requirement. The court agreed with the defendant that the court should act as a *Daubert*-type "gate keeper," even though the expert's testimony might be considered more technical than scientific. The court considered the expert's methodology in the light of peer review and/or publication, the known or potential rate of error, and the degree of acceptance within the relevant scientific community. The district court found that those factors weighed against the reliability of the expert's method; hence, it granted the motion to exclude the testimony.

The Eleventh Circuit reversed this decision on appeal, noting that the Supreme Court in *Daubert* had limited its holding to only those situations in which an expert relies on "the application of scientific principles" rather than "on skill or experience-based observation."

The Supreme Court held that a trial judge's general gatekeeping obligation applies not only to testimony based on "scientific" knowledge, but also to testimony based on "technical" and "other specialized" knowledge. It also held that a trial court may consider one or more of the specific factors that were mentioned in the *Daubert* decision; that the test of reliability is "flexible"; and that *Daubert's* list of factors may not apply specifically to all experts or in every case.

SUBSEQUENT CASES[11]

Joiner

In *General Electric v. Joiner*,[12] the respondent sued, alleging that his lung cancer had been "promoted" by his exposure to chemicals in his workplace that were manufactured or present on the petitioner's property. The petitioner moved for summary judgment, which was granted by the district court in part because the testimony of the respondent's expert had failed to show a link between the alleged exposure to chemicals and lung cancer and was therefore inadmissible. The court commented that the testimony was inadmissible because it did not rise above "subjective belief or unsupported speculation."

The Supreme Court reversed this decision and remanded the case. Its opinion was that it was within the discretion of the district court to conclude "that the

[11] Some of the factual information about these cases was gleaned from bvlibrary.com, a service of Business Valuation Resources, and an excellent source of valuation information.
[12] 522 US 136, 118 S Ct 512, 1997 U.S. LEXIS 7503 (Dec 15, 1997).

animal studies and the four epidemiological studies upon which the experts relied were not sufficient, whether individually or in combination, to support the experts' conclusions that the electrician's exposure to PCBs contributed to his cancer."[13] The Court commented further:

> But nothing in either *Daubert* or the Federal Rules of Evidence requires a District Court to admit opinion evidence which is connected to existing data only by the *ipse dixit*[[14]] of the expert. A court may conclude that there is simply too great an analytical gap between the data and the opinion proffered.[15]

Gross

In *Gross v. Commissioner*,[16] the issue in the Tax Court was the value of shares in a closely held S corporation transferred by gift. Both the respondent and the petitioners employed valuation experts. At one point in the trial, the petitioners moved to exclude the testimony of the respondent's expert because:

- His opinion of value was inadmissible because it was derived from the application of scientifically unreliable methodologies.
- The expert's underlying data and empirical analysis had not been published or otherwise submitted for peer review by the appraisal profession.
- Some of the data the expert relied upon was not available at the "as of date" of the valuation, and so buyers and sellers could not have relied on the information.

The court recognized that the expert testimony in the case was of a technical nature and accepted the gatekeeping role.

The court dispensed with the *Daubert* assertions in a fairly succinct manner:

- The discounted cash flow technique used by the respondent's expert was well known to the court and had been judged to be a proper methodology in a number of previous cases. The court therefore judged that the petitioners' argument was "nonsensical."

[13] *Id.*, 522, US at p.1.
[14] A dogmatic and unproven statement.
[15] *General Electric v. Joiner*, 522 US at p. 8.
[16] TC Memo 1999-254, 78 TCM (CCH) ¶ 201, TCM (RIA) 99254, (July 29, 1999).

- Both respondent's and petitioners' experts relied on the discounted cash flow methodology and therefore the issue of peer review was moot.
- The court accepted the assertion of the respondent's expert that it was valid to consider post valuation-date information because the underlying economics had not changed.

On this issue, the court decided that the testimony of both respondent and petitioner would stand as presented.

KW Plastics

In *KW Plastics v. United States Can Co.*,[17] there was a separate hearing on a motion by KW Plastics to exclude the expert testimony of a U.S. Can financial officer, who was testifying as to the quantification of lost profits and unjust enrichment. The court's role as gatekeeper was made difficult by the fact that the testimony of the U.S. Can expert was based largely on his experience rather than on specific analyses.

The court cited the Advisory Committee notes to Rule 702, which address the court's role in evaluating testimony that is based "solely or primarily on experience."[18] The court noted the Advisory Committee's comment that:

> If the witness is relying solely or primarily on experience, then the witness must explain how that experience leads to the conclusion reached, why that experience is a sufficient basis for the opinion, and how that experience is reliably applied to the facts. The trial court's gate keeper function requires more than simply taking the expert's word for it.[19]

The court, after examining the expert's report and more than 400 pages of deposition testimony, found that the expert's "calculations are speculative, without sufficient factual basis, and methodologically flawed. Furthermore, to any extent that the expert has used sound economic principles in determining U.S. Can's damages, the Court finds that its application of those methods to the particular facts of this case is impermissibly flawed."[20] The court therefore granted KW's motion to exclude the testimony.

[17] 2001 US Dist LEXIS 1630 (Feb 1, 2001).
[18] Fed R Evid 702, Advisory Committee Notes.
[19] *KW Plastics* at p. 4.
[20] *Id.* at p. 4.

The court further explained its position:

> The Court is not saying that an in-house employee like [the expert] cannot rely upon his own experience. However[, the expert] must exercise the same professionalism and thoroughness as any other expert in the field. U.S. Can has not persuaded the Court of the dubious proposition that a legitimate damages expert would not have sought or reviewed some type of extrinsic evidence to support his claims, particularly when that information was easily at his disposal.[21]

Cayuga Indian Nation

The case of *Cayuga Indian Nation v. Pataki*[22] is interesting in itself, but the *Daubert* issues therein are particularly applicable in the valuation arena. This case came to fruition after 18 years of litigation and unsuccessful negotiation attempts. The primary question was "how to compensate the Cayugas, in monetary terms, for the fact that through two separate transactions with the State they were dispossessed of their ancestral land in violation of the Indian Trade and Intercourse Act, and have remained out of possession of that land for the past 204 years."[23] The parties offered three real estate appraisers as their experts to testify:

- Expert 1 utilized a sales comparison approach.
- Expert 2 "employed a quantitative model which, in addition to real estate appraisal principles, relie[d] upon computer science, mathematics and statistics."[24]
- Expert 3 "offer[ed] yet another model; this one qualitative in nature, incorporating [appraisal principles and relying on, among other things, economics and history]."[25]

The *Daubert* hearing presented some difficult and unusual issues to the district court, which commented:

> As the three appraisers uniformly testified, what they do have in common is that despite over 90 years of real estate appraisal experi-

[21] *Id.* at p. 7.
[22] 83 F Supp 2d 318, 2000 US Dist LEXIS 761 (Jan 19, 2000).
[23] *Id.*, 83 F Supp 2d at p. 2.
[24] *Id.* at p. 2.
[25] *Id.* at p. 2.

ence among them, none has ever been confronted with an appraisal situation such as this. Thus, each of three appraisers was required to rely upon his own training, experience, and education to devise a valuation methodology which he believed workable for this unparalleled appraisal task.[26]

In its discussion of the *Daubert* issues, the court addressed:

- Qualifications of the experts
- *Daubert* principles
    - —Reliability
    - —Methodology
    - —Relevancy

Experts' Qualifications

The court held that all three experts were qualified to testify on the valuation issues. The court also observed that, during testimony, "it was implied that perhaps [Expert 3] was perhaps not qualified to testify because he was using economic theory and statistical analyses in his valuation."[27] After hearing testimony, however, and reviewing Expert 3's report, the court found "that although [Expert 3] is neither an economist nor a statistician, that does not mean that he is unqualified to rely upon those disciplines as part of his appraisal methodology in this case."[28] The court then pointed out that Expert 2 was not a mathematician, statistician, or computer scientist, but utilizing those disciplines in his valuation did not disqualify him from the case. The court also noted that complex problems often cannot be resolved without taking an interdisciplinary approach, and said that as long as a witness can demonstrate proficiency in the use of complementary disciplines, the use of such disciplines should not, by itself, disqualify the witness.

Daubert Principles

Reliability. The court concluded that the testimony of Expert 1 should be precluded because it failed to satisfy the reliability and relevancy considerations of *Daubert* and subsequent case law. In support of this decision, the court focused on the shortcomings in the methodology employed by Expert 1.

[26] *Id.* at p. 5.
[27] *Id.* at p. 3.
[28] *Id.* at p. 3.

Methodology. The court noted that although Expert 1 had used a recognized appraisal method (the sales comparison approach), his application of that method to the particular facts of the case was problematic. More specifically, the court was concerned about whether the comparison sales used by Expert 1 were truly representative and formed a proper basis for the sales comparison valuation method. The court was unsatisfied that Expert 1 and his assistants had complied with "established appraisal practices" in collecting and selecting the sales data relied upon, and noted that:

- There were discrepancies between the actual sales data and information recorded in Expert 1's analysis.
- Fair market values seemed to have been double-counted in some years.
- High and low sales data were discarded in some, but not all, of the years studied.
- What the court referred to as an "intuitive approach" was used to select comparable sales (i.e., Expert 1 testified that for each given year, he selected by "feel" the four sales to be used in his analysis, opining that "this is where the art of appraisal comes in").[29]
- Expert 1 did not always ensure that a given sale was an arm's-length transaction.
- Expert 1 failed to make any size adjustments in the sales data.
- Sales data from three to four years *after* the valuation date were considered.

In its concluding comments, the court noted:

> The valuation issues which this case presents are issues of first impression. That does not mean, however, that the testimony of [Ex-

[29] The court likened the means by which Expert 1 selected the comparable sales to the "subjectiveness" of the methodology of the expert in *Kumho*. The court went on to comment:

> The Court is not saying that subjective analysis has no place in the valuation task which these appraisers were facing. Indeed, at least from what the Court was able to glean from the hearing testimony of Experts 2 and 3, they too, necessarily, relied upon a certain amount of subjectivity, particularly with respect to deriving fair market value for pre-1900 sales. One primary difference, though, between the testimony of [Experts 2 and 3] and that of [Expert 1] is that the latter continually relied upon little more than his subjective "feeling," especially in terms of selecting sales to be used in arriving at a fair market value for any given year. [Experts 2 and 3], on the other hand, both have significant, reliable objective data, which they then relied upon to extrapolate their conclusions.

Id. at p. 5.

> perts 2 and 3,] who both freely admitted to developing their respective valuation methodologies for first-time use in this case, should be precluded from testifying. The Court finds no basis for precluding the suggested appraisal methods of either [Expert 2 or 3] because, in its opinion, at least at this juncture, both methodologies are reliable and relevant.[30]

Relevancy. The court found that the testimony of Expert 1 "would not have been helpful to the jury in understanding or determining how the subject property should be valued."[31] It stressed that this was not because of Expert 1's conclusions, but rather because of the manner in which his chosen methodology was applied.

It is interesting to note that Expert 1 apparently testified that his analysis of each sale was not complete, because of "time and budget constraints." The court expressed sympathy with that situation, but noted that Experts 2 and 3 were able to develop formulas and present acceptable data, even though they operated under similar constraints.

Target Market Publishing

In *Target Market Publishing, Inc. v. ADVO, Inc.*,[32] the issue was lost profit damages. Target and ADVO had entered into a joint venture that failed; Target alleged that ADVO had not fulfilled its obligations, and thus Target sought damages measured by lost profits. Target relied on an expert report prepared by an accountant and business appraiser, who concluded that "had ADVO performed its obligations under the Joint Venture Agreement, [Target] should have earned $1.4 million during the contract period ending 7/31/94."[33] ADVO's position was that this report was "pure speculation, based upon utterly implausible assumptions and unreliable methodology."[34] Although there apparently was no mention of *Daubert* or any separate *Daubert* hearings, the district court seemed to agree with ADVO's position, stating that Target "relies upon mere assumptions . . . from which no reasonable inference of lost profits could be drawn."[35] The district court added that "the entire body of evidence that is not mere speculation does not support an award of lost profits that satisfies the jurisdictional minimum."[36]

[30] *Id. at p. 10.*
[31] *Id.* at p. 9.
[32] 136 F3d 1139, 1998 US App LEXIS 2412 (Feb 18, 1998).
[33] *Id.*, 136 F3d at p. 2.
[34] *Id.* at p. 2.
[35] *Id.* at p. 2.
[36] *Id.* at p. 4.

The circuit court concluded that the district court did have *Daubert* in mind and had decided to exclude the expert's report and testimony on those grounds. The circuit court affirmed this decision.

The circuit court commented on the expert testimony in light of its decision and noted that the projection of $1.4 million in profits was based "upon assumptions that do not legitimately support the conclusion."[37] As an example, the projected profits assumed that the joint venture business would achieve full penetration into dozens of markets in which no sales effort had even begun, although the joint venture agreement was more than half completed. This conclusion flew in the face of the fact that in the first market attempted, the level of business had declined steadily. The forecast was also based on the assumption that the targeted advertisers would pay full price for their ads, though this was, apparently, not what could reasonably be expected to happen.

Lessons Learned

It is clear that the courts are going to take their gatekeeping duties seriously and that *Daubert* challenges will become a regular, and perhaps extensive, part of litigation involving experts. It has also been suggested that *Daubert* challenges may be used to discover additional information about or from an adversary expert.

In the early days following the *Daubert* decision, we might have assumed that appraisers, accountants, and economists were not included and that the standards to be used in evaluations of experts were rather rigid. It now is clear that *all* experts are subject to *Daubert* examination and that the courts have flexibility in the application of standards.

We observe that it can be acceptable to modify methodologies to address new or unusual situations, as long as the expert is prepared to explain the modified procedures and the need for them. We also observe that the employment of complementary disciplines (outside of the expert's core or titular expertise) can be acceptable, again so long as the expert has at least some experience in the use of these disciplines and can satisfactorily explain what was done. It is also clear that expert testimony based only on experience will be very difficult to get past the courtroom doors.

Experts must now be prepared to discuss their methodologies in some detail before the start of their work—perhaps even before they are retained. That is, the attorneys directing litigation will want to satisfy themselves early on that prospective experts will pass muster in a *Daubert* challenge. They will want to know how a prospective expert plans to proceed, and are well advised to do

[37] *Target Market Publishing, Inc. v. ADVO, Inc.*, 136 F3d at p. 4.

some independent checking of the methodologies proposed. These decisions may even determine whether or not to litigate. This can be a challenge for experts, who know full well that methodologies may change midstream in response to the character and quantity of data that turns out to be obtainable. As an example, prelitigation discussions may assume that certain data about adversary operations will be available or that certain market information can be gleaned. If these assumptions do not materialize, the expert may have to use surrogate data and rely more heavily on experiential testimony that might not withstand a *Daubert* challenge.

It seems to us that the courts are going to have "scientific method" in mind when they address *Daubert* issues, at least to some extent. It behooves us, then, to think in these terms as well, as they relate to what we do as experts.

PASSING MUSTER AFTER *DAUBERT*

Qualifications

The Federal Rules of Evidence and the case law show that an expert's professional qualifications are an important ingredient in passing the *Daubert* test. We note, however, that membership in specific professional organizations is not by itself an essential ingredient in that qualification. This is probably as it should be, because it gives courts some flexibility in evaluating the qualifications of potential expert witnesses and allows them to accept a witness who has gained expertise through experience.

In the valuation field, there are standard-setting organizations, and there are independently published sets of standards, as well as organizations that offer both tested professional qualifications *and* professional standards. Some government agencies that frequently deal with valuation issues have also made pronouncements about standards and methods. It is beyond our scope to offer a detailed description of each of these sources here,[38] but it is useful to note their existence:

- The Appraisal Foundation
- Uniform Standards of Professional Appraisal Practice (USPAP)
- Internal Revenue Service

[38] For a more extensive description of these organizations, the reader should consult Shannon P. Pratt, Robert F. Reilly, and Robert P. Schweihs, *Valuing a Business*, 4th ed. (New York: McGraw-Hill, 2000).

- Department of Labor
- Association for Investment Management and Research
- ESOP Association
- American Society of Appraisers
- Institute of Business Appraisers
- National Association of Certified Valuation Analysts
- American Institute of Certified Public Accountants
- The Canadian Institute of Chartered Business Valuators
- American Institute of Real Estate Appraisers

Scientific Method

There was a sudden upsurge of interest in the so-called scientific method when the Supreme Court interpreted Rule 702's requirement that expert testimony pertain to "scientific ... knowledge" to mean that the expert propounding the testimony must have "a grounding in science's methods and procedures."[39]

There are a number of versions of the scientific method but it appears to us that, in general, it is an iterative process involving:

- Hypothesis development,
- Hypothesis testing,
- Hypothesis confirmation, and
- Publication and peer review.

If a hypothesis is not confirmed by testing, then a new hypothesis may be developed and the procedures started over again.

Hypothesis Development

A *hypothesis* is a clear statement, which might also be called an educated guess, about cause and effect in a given situation (e.g., "radio interference is caused by sun spots"). The hypothesis should be testable or, to use the words of *Daubert* and other cases, "falsifiable" (i.e., capable of being proved false). A proper hy-

[39] *Daubert*, 509 US, p. 2, syllabus (b).

pothesis is one that can be empirically tested; tests must be designed that will (1) give a clear yes-or-no answer to the hypothetical question (the premise of the hypothesis) and (2) rule out other hypotheses.

Hypothesis Testing

In the field of physical science, testing often takes the form of repeatable experimentation. In the social sciences, hypothesis testing may be conducted through and based on interviews, observations, and surveys. Protocols are usually developed to guide the collection and organization of information so that the testing will not be guided along the way by the experimenters.

Obviously, the testing phase is extremely important and it is here that the reliability of the methodology can go awry:

- The observer simply omits or limits testing after only a few tests, concluding either that the hypothesis has been confirmed or that it does not appear to require testing because it appeals to common sense or logic.
- The observer is biased as to the choice of tests or their outcome.
- The observer chooses to ignore or rule out data that does not support the hypothesis.
- The observer simply makes errors in the observations or the analysis of the observations.
- The observer fails to use a possible triangulation approach (that is, to use more than one possible testing methodology) to observe whether the results converge.

Hypothesis Confirmation

In this step the scientist must carefully examine whether or not the test results confirm the hypothesis. The scientist must address a number of questions, such as:

- Were the tests accomplished error free?
- Do the test results clearly confirm or disprove the hypothesis?
- Are the test results unambiguous?
- Were the tests adequate (i.e., is there a reasonable possibility that another untested phenomenon could explain the hypothesis)?
- Was there adequate cross-checking and triangulation?

Peer Review

It is common for scientific results to be published in journals that are circulated to those experienced in the field. Scientific papers may be reviewed by a panel of experts or the editors of the journal (or both). In either case, the scientific papers published are subjected to public criticism by the journal readers. Prestigious journals publish only a small portion of the papers submitted and typically are rigorous in their evaluation of the materials submitted.

Scientific papers are often presented at meetings of professional societies or associations. Although there may be some evaluation of the material before it is accepted for presentation on such a program, the audience reaction is largely undocumented and therefore of limited use as true peer review.

Eventually, scientific conclusions, if they survive the peer review process, find their way into textbooks and become part of what are considered authoritative sources.

VALUATION AND THE SCIENTIFIC METHOD

If we accept that a valuation or damage quantification requires technical and/or specialized knowledge, then it is useful to observe the extent to which those efforts fit under the rubric of the scientific method.

Hypothesis Development

A formal hypothesis is not typically enunciated in the preparation of a valuation opinion or damages quantification. A professional with a significant amount of experience, however, likely has mentally made an educated guess about what an analysis will conclude or where a valuation will come out, if only on a very rough basis. This guesstimate might be based on the expert's experience in similar situations, some rough calculations, and/or the opinions of the client or others. Whatever its genesis, our experience suggests that a working hypothesis usually is conceived at the start of a valuation assignment, whether or not it is enunciated.

Examples of such working hypotheses might include:

- The market value of the subject property is zero.
- The market value of the subject property is $1 million (i.e., some "gut-feeling" number).
- Specific buyer A should be willing to pay $500,000 for the subject property.

- The alleged infringer profited from use of the intellectual property (IP).
- The IP owner suffered economically from the alleged infringement.
- An arm's-length license between the parties would have contained a 5 percent royalty (again, a viscerally based guess, albeit based on experience and industry custom).

Because hypothesis testing is an iterative process, it does not matter whether the valuer's initial thoughts are close to the final conclusion. If testing shows that the initial assumption has no validity, it is, in effect, abandoned and a new hypothesis developed. The analysis will eventually center on a hypothesis that is confirmed by an expert's tests, no matter what the initial starting point. In practice, this is a continuous process without observable stops and starts. Also, as we will show in the following paragraphs, there is a fairly limited range of test methods, so the importance of the original and subsequent hypotheses fades during the process.

It may be possible that a valuation or damages expert is not qualified to test a particular hypothesis, such as:

- The subject property is indisputably owned by one of the parties.
- The ownership of the subject property can, in fact, be transferred.
- The legal rights to the subject intellectual property are valid.

The expert may therefore simply accept such hypotheses as true, as part of a set of assumptions under which the analysis is done. An expert is sometimes requested to assume facts at odds with current or typical conditions, such as an "as of" date in the future or past. This is not a problem as long as those facts are clearly declared as an input to the analysis.

Hypothesis Testing

The valuation or damages professional tests a hypothesis by estimating the economic result of a surrogate transaction or by investigating the economic result of actual transactions (in the case of owner's damages or infringer's profits).

The Virtual Transaction

We characterize this process as the use of a "virtual transaction." This term is not commonly used by valuers. More typically, they list a series of underlying assumptions that guided their work. In effect, however, they are actually describing a virtual transaction, and so we find it useful to use that terminology and description in this discussion.

When we opine on the value of intellectual property or a royalty for its use, we are describing the economic results of a transaction that has not and will not take place. If, as an example, our assignment is to estimate the market value of a trademark or opine about an arm's-length royalty that should be paid for its use, we are not able to put the trademark on the market and observe its selling price, nor can we negotiate a license agreement in order to observe the resulting royalty rate. The owner of the mark likely has no intention of selling or licensing it, but needs the valuation or royalty opinion for another purpose. We therefore must simulate the relevant transaction using methods that are used in real life by real buyers and sellers as they negotiate prices and royalty rates.

This is a two-stage process in which we (1) describe the virtual transaction; and (2) describe the most probable economic results of the virtual transaction.

Description

The first step in the analysis is to carefully describe the virtual transaction, because this description will guide the hypothesis testing.

We must make some decisions about the relevant characteristics of the transaction we are trying to simulate. For example, we have to consider whether we are trying to replicate market value (any buyer or seller), or investment value (special buyer and/or seller). We also have to consider whether buyer and seller are willing to make the transaction or whether there is some degree of compulsion. We need not identify specific buyers and sellers, but we do have to characterize them so that it is clear what kind of a virtual transaction is contemplated.

The timing of the transaction must be delimited: is it current, in the future, or at some time in the past? Value changes with time because the conditions that drive value are continually changing.

Valuations are forward-looking; that is, the value of property rights as of a given date rests on the value of the economic benefits to be received after that date. We are, however, interested in the relevant history of an asset, particularly its economic performance. This history sets the stage for our forecasts of the future and helps us to better understand the economic performance of the asset at the appraisal date. When we quantify damages, we are focused more on history; the future is largely irrelevant because the litigation giving rise to the need for damage quantification is expected to resolve the issues. Damages may, however, be based on the degradation in value that has occurred due to past actions. In such a situation, future time periods become relevant, because valuations act as the basis for the damage measurement.

It is also essential that we understand the property rights that are the subject of the study. Are we focused on the entire bundle of property rights, or some part of it? This is not always intuitively obvious. We need to know the reason *why* the value or damages quantification is being sought, as the reason itself may dictate other conditions of the virtual transaction.

All these elements are discussed at greater length in the chapters of the main book.

The initial assumptions about a virtual transaction are important to the reliability of the valuer's conclusion. Assume, as an example, that the assignment is to estimate the market value of oceanfront land at a New Jersey resort. It would be a serious error, in describing the virtual transaction, to assume that the buyer intended to use the land for farming. Such a buyer might be delighted to pay a price based on farm acreage, but a seller would never sell at that price, so the described virtual transaction does not comport with reality. It is therefore useless in the testing of the value hypothesis in this case.[40]

A complete description of the virtual transaction is also essential to enable the reader of a valuation report to understand the conclusion. For example, an assumption of a virtual sale between willing parties is going to produce quite a different price from a scenario in which one of the parties is compelled by circumstances to make the deal. Without that information, one cannot judge the validity of the value opinion.

The answers to these real-life buyer/seller questions are difficult, and each one can spawn a myriad of additional questions. Nevertheless, the answers—as long as they are based on real-life data—will help us confirm or discard the working hypothesis.

Economic Result of the Transaction

When the virtual transaction has been described, we can proceed to estimate its economic result. In valuation, we can quantify the economic result of the virtual transaction using the three traditional methods:

- Cost approach
- Market approach
- Income approach

[40] This relates to the concept of "highest and best use" that is commonly referred to in the valuation of real property. It is assumed, in the virtual transaction, that buyer and seller both know the most efficacious use of the property and make the transaction on that basis.

These are based on economic reality and investment theory, as fully described in Chapter 6.[41] These methods and principles are also based on the actions of real-life buyers and sellers thinking to themselves:

- I can either buy this property or create one just like it, and what would that cost? (cost approach)
- Other people have bought this kind of property; what did they pay? (market approach)
- What am I justified in investing for this property given the money I can make with it in the future? (income approach)

The same methods, questions, and principles guide an investigation directed at royalties or damages due to lost profits or unjust enrichment:

- What would it cost to obtain the economic benefit of the licensed property by alternate means? (cost approach)
- What have others charged as a royalty for the use of a comparable property? (market approach)
- What royalty amount represents a reasonable division, between owner and licensee, of the economic benefit from using this intellectual property? (income approach)
- How would sales revenues, expenses, and profits be affected if the subject technology or patents were infringed? (but-for accounting analysis)
- How would sales revenues, expenses, and profits be affected if one had free access to the subject technology or patent? (but-for accounting analysis)

Figure 11A.1 illustrates the process of using the virtual sale transaction to estimate market value. (For simplicity, only the income approach is illustrated.) This chart should be read from the bottom, where the virtual transaction is described. In the center we describe some of the data-gathering and analysis tasks that are dictated by either the method or the nature of the virtual transaction.[42] The objective of performing these tasks is to obtain the ingredients needed for

[41] We note that there are many techniques for developing and analyzing the elements (i.e., forecasts, discount rates, and financial market data) required by these three basic methods. These are tools to develop the ingredients, however, not basic methods themselves.
[42] This is not intended to be an exhaustive list of such tasks.

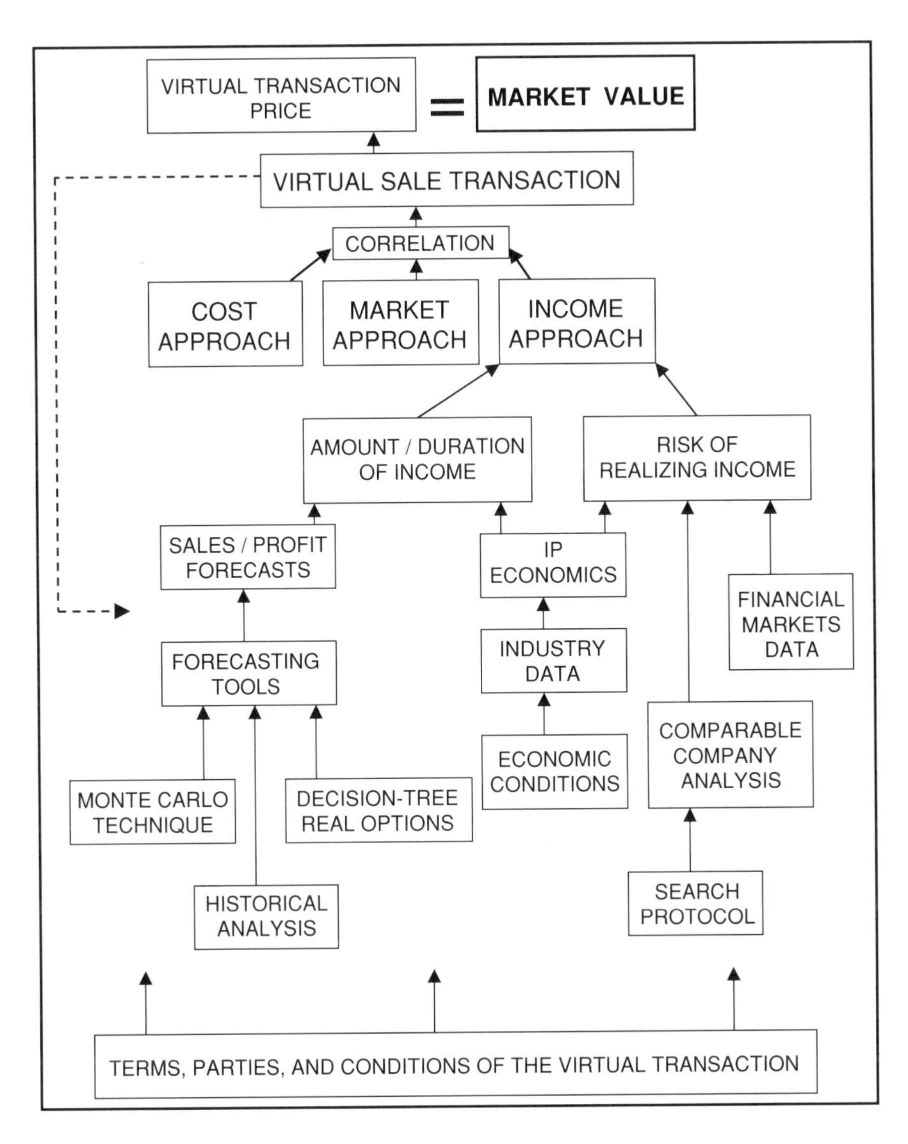

Figure 11A.1 Analyzing the Virtual Sale

(in this case) an income approach calculation. There may, in fact, be more than one income approach calculation.

The process then requires a correlation of the estimates produced by the three methodologies. In our discussion of the scientific method, this was described as *triangulation*. This is when the expert reviews the data and analyses used throughout the study and reconciles the results into a single opinion about the "price" that would result from the virtual transaction. That price is equivalent to market value. Anomalies that arise in the correlation step may cause the expert to revisit previous data-gathering and analysis results, as indicated by the dotted line in Figure 11A.1.

These, then, are the general steps that one would follow in application of the scientific method to develop an opinion about market value.

Arm's-Length Royalty

When the engagement involves providing an opinion of an arm's-length royalty, the tasks are not unlike those previously described.

- There are cost approaches: What royalty will provide a return of and return on my investment in this intellectual property (licensor)? What royalty am I justified in paying relative to the cost of using alternative intellectual property (licensee)?
- There is a market approach: What royalties are others paying in arm's-length transactions for similar intellectual property?
- There is an income approach: What will be the economic benefit of exploiting this intellectual property, and how should that be divided between licensor and licensee?

Each of these approaches requires data-gathering and analysis tasks.

The virtual transaction is some form of license, and once again we have the essential task of carefully describing that transaction, to enable understanding and validation of the royalty opinion. The data-gathering tasks may be quite similar to those undertaken in a valuation, though some are quite different. Again, the results yielded by different methods should be correlated in coming to a conclusion about the royalty.

Lost Profits/Unjust Enrichment Damages

The scientific method parallels the steps just described, with the most notable exception being the use of actual transactions as a benchmark. Again, there is a series of virtual transactions, the effect of which is measured by comparing them to the actual. If we can assume that the parties to the dispute are businesses that that sell goods or services, our task is to observe the economic result of the following transactions:

- The plaintiff's sales of goods or services.
- The defendant's sales of goods or services.

The benchmark is established by observing the economic result (profit) of the *actual* sales of goods or services by both parties. We then observe the economic result (profit) of a series of virtual transactions (plaintiff and defendant sales of goods or services unaffected by the disputed activity).

By comparing the results of the actual with the virtual, we measure the economic result of the disputed activity:

- Did the plaintiff suffer reduced profits? (fewer sales, degraded prices, increased expenses)
- Did the defendant enjoy enhanced profits? (greater sales, higher prices, reduced expenses)
- Did both of the preceding occur?

As noted earlier, both the actual and virtual transactions should be carefully described,[43] data must be gathered concerning the external and internal factors that affect sales and profitability, and a logical analysis must be developed to quantify the actual and virtual sales.

WHAT CAN GO WRONG?

Describing the Virtual Transaction

We have previously stressed the importance of describing the virtual transaction from which the conclusion will be extracted. The reason for this emphasis is that the description not only characterizes the ultimate conclusion, but also guides the data gathering and analysis. If the objective is to estimate the *market value* of a semiconductor patent on *July 31, 2001*, those two pieces of information, at the very least, must be included as descriptors of the virtual transaction, especially in a case in which the critical date is something other than the current one. If this

[43] The description should include such matters as: Who are the buyers and sellers? What are the relevant products or services? What is the proper time period? and so forth.

information is not used in the virtual transaction, the conclusion drawn from it will be meaningless, no matter how astute the analysis.

The essential question is whether the chosen description of the virtual transaction aims the analysis in the right direction, so that a proper analysis will land the conclusion on target. For example, if the virtual transaction is between a willing buyer and willing seller, the market value hypothesis testing should include an investigation into what would influence the actions of each party. We might therefore investigate general economic conditions, conditions in a particular industry or region, and the financial markets. If, in contrast, the virtual transaction involves a specific buyer or class of buyers, the investigation will be much more focused and may include elements that would not even be considered in the first example.

In general, the description of a virtual transaction includes:

- A description of the subject property and the rights being transferred
- A description and characterization of the buyer and seller (premise of value)
- The "as of" date
- The intended use of the analysis
- Whether the subject property is to be transferred by itself or along with other assets

Analysis Tasks

In the gathering and analysis of data, the scientific method would have us make exhaustive searches and analyze the accumulated data in unbiased and rigorous fashion. All experts would agree with this objective.

This perfect objective is not unlike our expectations of the news media, and members of that profession would likely agree unanimously. There are, however, real-life impediments to the realization of this perfection. The media must "boil down" a complicated event into several paragraphs or a two-minute communication. Is it possible to do this without being influenced by human feelings and outlooks? Probably not, although if the analysis involves several people, their individualities might cancel each other out. Scientists and valuers face the same problem, and the solutions are not necessarily easy. It is impossible for an intellectual property expert to discover and analyze the entire universe of facts about a given situation. The process of selection, however necessary, is critical.

The reliability of the selection process and performance of the analysis tasks can be subtly affected by relationship pressures or by inadvertent error, oversight, or laxity. The following highlights some of these tasks and points out some pitfalls.

Relationship Pressures

A valuation or estimate of damages in the litigation context involves many parties, including the expert(s), client(s), and attorneys. Each have their own objectives. The expert's objective is to perform a competent analysis, form an unbiased opinion, and be an advocate for that opinion. The other parties are permitted a bit more latitude in their advocacy, and will argue from their particular self-interests to affect the outcome. Experts can also become the target of those arguments, so they must be vigilant to deflect those pressures that might adversely affect the analysis or bias the opinion.

Obviously, the most visible relationships involve economic benefit.[44] Thus, professional fees that are contingent on the outcome of the analysis are clearly prohibited. Contingency can take many forms, however, even including legitimate fees unpaid or withheld at the time expert testimony is rendered.

Relationship pressures can affect "pure" scientific research when the agency sponsoring the research has a financial interest in the outcome. Most experts prefer challenging engagements and value relationships with clients or other professionals (such as attorneys) who provide such opportunities. Experts do not wish to disappoint such sources. Experts exist on their professional reputations and welcome opportunities to be associated with a landmark case[45] or to move into new areas of expertise or specialization. These situations can introduce pressures that may not exist in more ordinary client relationships.

Self-Evident Facts

Some facts may appear to be self-evident, and an expert may thus decide that exogenous support is not necessary. As an example, "everyone knows that money market rates are less than 1 percent." Such facts—which are really conclusions—should be carefully examined and considered.

Guided Fact Gathering

Fact gathering can be "guided" by a client in many ways, both subtle and other-

[44] There is little new about this. A seventeenth-century artist, existing on the patronage of nobility, could be excused if he rendered family portraits rather sympathetically. In this situation, the influence on the artist's "opinion" resulted from economic pressures that could have been considerable.
[45] Such a case may be large (in terms of dollars at risk), precedent-setting, or involve a celebrity client or adversary.

wise: "we will put you in touch with some of our customers" or "we can copy some articles on this subject for you" are some of the former. Nothing is wrong with this per se, but perhaps what these customers or articles say should be checked against other sources, or perhaps the selection of customers or articles for research should be made by the expert.

Limited Fact Gathering

Fact gathering might be limited to data that is easily at hand: "We did that search in a former case, let's dig that out and use it." Again, nothing is wrong with that, but if the data is likely to be affected by the passage of time or changes in other conditions, it should be rechecked. Things change, new data emerges, old facts get discredited (or confirmed).

Lack of Search Protocol

Having no search protocol can lead one to collect only the "low-hanging fruit"—that is, to go with what can be obtained easily. A search or selection protocol is important because it may eliminate at least one level of bias. We must recognize, however, that the protocol itself can influence, or even mandate, a result. For example, at one time a survey of public opinion by telephone was deemed suspect because of the skewed population of telephone subscribers. That is no longer true, because telephones are ubiquitous in U.S. homes, but reliance on Internet surveys revives the old suspicions.

Designing a search protocol helps to ensure that the data gathered will be relevant to the question at hand. One can therefore have more confidence that conclusions based on the factual evidence of the search are reliable. A search protocol would include such elements as:

- The universe of information to be searched;
- The standard for extracting a subset of relevant information;
- The elements of information to be recorded; and
- How the search will be documented to permit verification and testing.

In a valuation or damages example, if one were seeking data about one segment of the paper industry, one method might be to select a group of companies to examine. The search protocol might be to search the SEC database of company filings, and to extract the most current 10-K report for:

- Companies with common stock, publicly traded on a U.S. market,
- With an SIC code of 2621 (paper mills),

- With total sales revenue between $8 and $15 billion,
- With at least 75 percent of sales revenue from paper mill operations,
- With positive net profit in the current year.

This is a discoverable and verifiable population of companies; the existence of the protocol means that the researcher (at least at the beginning of the process) does not have the freedom to pick and choose. There can still be argument about whether the protocol itself is appropriate, but at least the arguable facts are presented.

There are times when the search protocol may constitute a search *method.* Assume, for example, that we are trying to discover a company's selling price of widgets during the past two years. We could, as an example:

- Accept the data from all the many thousands of invoices.
- Select a sample of invoices using invoice number and a random number generator.
- Select a sample of invoices, choosing every *nth* invoice.

By the use of a search protocol, the researcher is not influencing which data is selected for analysis. However, there are situations in which no search or selection protocol can be used—other than "we will make a diligent search for information relating to X and examine everything we find."

Overly Generalized Search

The generalized search is a more dangerous minefield for the expert because there is more opportunity to be influenced by one's feelings.

- Will an Internet, keyword search bring up all the relevant sources? We may have more confidence than we should that the answer is yes. Not all search engines are created equal, and not all worthwhile data is even Internet-accessible.
- If the search, Internet or otherwise, turns up what appears to be the mother lode of information we are seeking, we may give in to the natural inclination to stop searching. If, for example, we are searching for information about the paper mill business, a logical place to go might be a trade association. We should be aware, however, that information from such a source might provide only good news about the industry. Whatever the source, we must be sensitive as to possible bias, slant, or "spin" in the information we are observing.

- An expert's searches should be documented, to include a description of what potential sources were searched and what was found (and not found). How else can one defend one's methods? This may be an unfamiliar requirement for some valuers or intellectual property experts.

The authors were involved in a *Daubert*-type proceeding in which an expert, opining on an arm's-length royalty rate, based his opinion in large part on "interviews with influential persons in the industry." These talks took place as a part of the expert's conduct of his practice. He was unable, however, to provide the court with the names of the interviewees, the dates when the interviews took place, or the specific subject matter of the discussions. Three strikes and he was out.

- Under Rule 26 of the Federal Rules of Civil Procedure, an expert witness is expected to prepare a written report that includes:

  —A statement of all opinions to be expressed

  —The basis and reasons therefore

  —Data considered by the expert in forming the opinions

  —Exhibits to be used in support of the opinions

  —Qualifications of the expert

  —A list of publications authored by the witness in the preceding 10 years

  —A list of testimony or deposition appearances during the preceding four years

  —Compensation to be paid the expert.

Just as with the principles set forth in *Daubert*, it is not unreasonable to expect that these federal standards will slowly permeate the courts at all levels.

Economic Result of the Virtual Transaction

Having gathered the available data in a comprehensive and unbiased fashion, an expert comes to a conclusion *based* on the data analyzed and relied on. There should be no unexplained disconnect between the facts and the conclusion. Therefore, an expert should make clear what data was relied on and what data was discarded, and the reasons therefore, so that there is a clear and understandable path from fact to conclusion. There are many reasons why disconnect problems arise:

- An expert may have prepared or testified to a position in a past, similar case and feel considerable pressure to be consistent with that position, even if the fact pattern in the instant case does not support it as cleanly or completely. Some experts may be uncomfortable in admitting that (and explaining how) their opinions have changed over time or as the result of new studies or analyses.
- An expert may cling to a position because others in his or her firm have espoused it.
- An expert may cling to a position because he or she is tied to an early and unconsidered conclusion communicated to the client that is now (after analysis) poorly supported by the facts.
- An expert may have followed procedures by rote and not become aware of or accounted for subtle differences in the instant fact pattern.
- A "black box" may have been used in the analysis, in the form of a computer model with a built-in bias. This can take the form of a simple oversight, such as using a computer spreadsheet from a previous engagement that has some embedded but nonobvious factors that are inappropriate for the instant case. An expert may simply not know what is in a black-box computer calculation or program he or she is using.
- One of the *Daubert* questions relates to whether the "rate of error" can be discerned or is known. This is a difficult test in the valuation field, because little of what we do is based on statistical analysis. Our objective is most often a single opinion rather than a range of opinions. We can, however, demonstrate the effect on value of a range of reasonable underlying assumptions. This could be a useful step if this question is likely to be addressed. At the least, an expert should be prepared to discuss why such a sensitivity analysis is or is not appropriate.

Although there is not a plethora of authoritative sources for valuers and damages experts, some are available. This book is one, and its footnotes and Appendix F list others. Many essential elements of the valuation process—such as the cost, market, and income approaches; premises of value; and the derivation of discount and capitalization rates—have been largely accepted by experts, litigants, and the courts as "conventional and uncontroversial wisdom." We may, however, have reached a point at which this assumption can no longer be made; even conventional wisdom may have to be proven if challenged.

Courts may also begin to seek peer-review evidence concerning some elements of valuation opinions that heretofore have been routinely accepted. Though perhaps not as rigorous as the reviews done by some scientific organizations, appraisal professional groups do provide a level of peer review relative to valuation concepts.[46] Experts should seek these out.

SUMMARY

We have discussed a number of potential pitfalls facing valuation and damages experts relative to intellectual property. Obviously, simple errors are another one, but these really do not deserve separate discussion. Rather, we have attempted to frame the discussion using the scientific method as a guide, because that may well be the framework of a *Daubert*-like review.

We are seeking to develop expert testimony that will pass the *Daubert* tests; that is, expert opinions that:

- Are based upon sufficient facts and data.
- Are the product of reliable principles and methods.
- Represent a reliable and appropriate application of those principles and methods to the facts of the case.
- Analyze the data in a manner that can be tested, when this is possible and appropriate.
- Uses methods that have been subjected to peer review and publication.
- Is not merely subjective belief or unsupported speculation.
- Uses methods for which a rate of error can be known.[47]
- Employs data and analysis methods that are accepted in the relevant community.

We hope that pointing out some of the factors that may cause shortcomings will assist experts in avoiding them.

[46] The American Society of Appraisers, 555 Herndon Parkway, Suite 125, Herndon, VA 20170 is such an organization, and its Business Valuation members have been especially active in examining and documenting valuation practices.

[47] Although this is a "*Daubert* principle," it is unlikely that it is applicable to valuation or damages measurement except when specific statistical studies might have been employed.

14A (New)

Trademark Dilution—A Discussion of Damages and Valuation Theory

The main chapters of this book do not discuss issues of intellectual property damages; they are focused more on compensatory payment rather than property value. An issue has arisen, however, in the area of trademark litigation that prompts a consideration of damage theory and its basis in valuation.

INTRODUCTION

The Federal Trademark Dilution Act of 1995 (FTDA) became effective January 16, 1996. It provides that:

> The owner of a famous mark shall be entitled, subject to the principles of equity and upon such terms as the court deems reasonable, to an injunction against another person's commercial use in commerce of a mark or trade name, if such use begins after the mark becomes famous and causes dilution of the distinctive quality of the famous mark, and to obtain such other relief as is provided in this subsection. In determining whether a mark is distinctive and famous, a court may consider factors such as, but not limited to—

(A) the degree of inherent or acquired distinctiveness of the mark;

(B) the duration and extent of use of the mark in connection with the goods or services with which the mark is used;

(C) the duration and extent of advertising and publicity of the mark;

(D) the geographical extent of the trading area in which the mark is used;

(E) the channels of trade for the goods or services with which the mark is used;

(F) the degree of recognition of the mark in the trading areas and channels of trade of the mark's owners and the person against whom the injunction is sought;

(G) the nature and extent of the use of the same or similar marks by third parties; and

(H) the existence of a registration under the Act of March 3, 1881, or the Act of February 20, 1905, or on the principal register.[1]

It appears that the just-named factors apply more to the question of whether a mark is "famous" than they do to determining whether a "dilution of the distinctive quality of the mark" has taken place. The absence of a clear direction in this regard has given rise to some disagreement among the courts and some frustration among litigants. In this chapter we do not address the question of what constitutes a "famous" mark because the answer to that question is unlikely to depend on valuation or economic issues. We do, however, address the question of how to detect "dilution" by introducing additional factors based on valuation and economic principles.

BACKGROUND

Collisions of trademarks are occurring more frequently than in the past as trademark owners explore new avenues for their exploitation. As an example, Indian Motorcycles of Gilroy, California, and the Cleveland Indians Major League Baseball team are in litigation over the use of the term INDIANS and the motorcycle company's INDIAN logo for clothing and toys.[2] Neither entity is in the business of manufacturing and marketing clothing and toys, but each has or intends to extend its marks to these secondary exploitations, presumably through

[1] 15 U.S.C. 1125, (c)(1).
[2] "Legal Battle over Indians," *Mercury News*, June 25, 2002, as reported in the *INTA Bulletin*, August 1, 2002, p. 10.

licensing. These marks have coexisted for years but now collide. As businesses expand, the geographical separation of markets is also breaking down, leading to more collisions and "near misses" involving trademark rights. These issues are likely to arise with increasing frequency.

When trademark collisions occur, the thought process of the plaintiff might evolve as shown in Figure 14A.1.

The form of infringement called counterfeiting is the most obvious and clear to identify, since a counterfeit is defined as "a spurious designation that is identical with, or substantially indistinguishable from a designation"-such as a mark registered at the U.S. Patent and Trademark Office.[3]

[3] 15 U.S.C. 1116, (d)(1)(B)(i) and (ii).

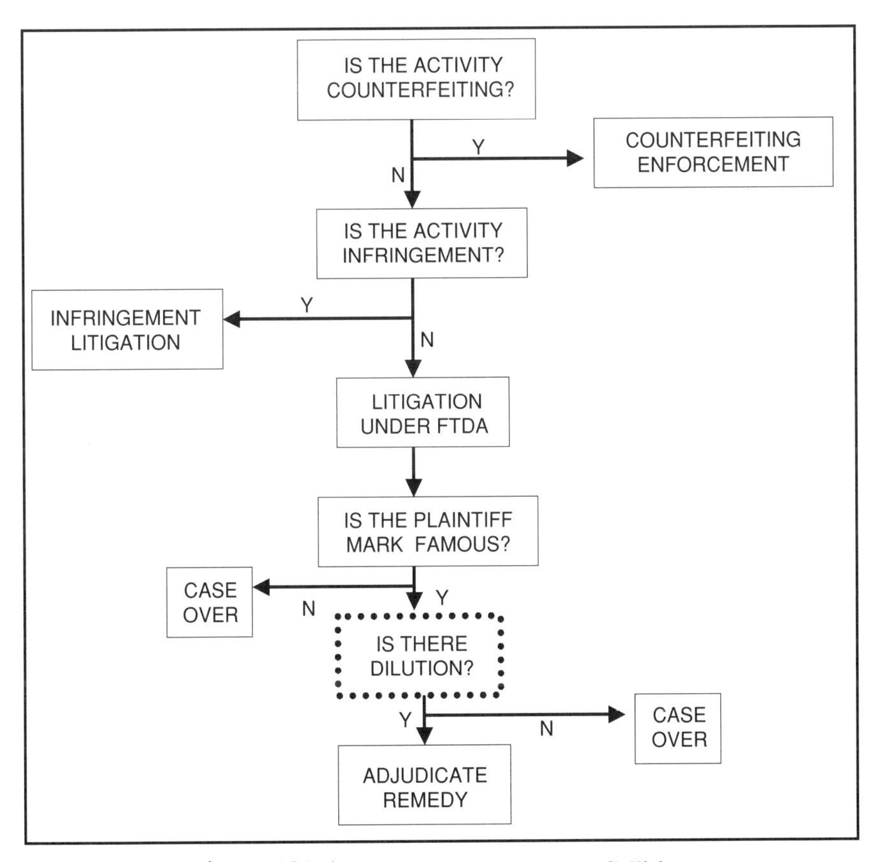

Figure 14A.1 **Analyzing a Trademark Collision**

Infringement, other than counterfeiting, is often less obvious and more difficult to identify since it involves "use in commerce [of] any reproduction, counterfeit, copy, or colorable imitation of a registered mark . . . [in] which such use is likely to cause confusion, or to cause mistake, or to deceive."[4] Identifying infringement involves more than just a comparison of nearly (or exactly) *identical* marks but a judgment as to whether *similar* marks cause confusion in the marketplace as to origin.

It seems to us that the least obvious trademark collision and the one most difficult to identify is that which is described in the FTDA as a situation in which "such use . . . causes dilution of the distinctive quality of the famous mark."[5] Even more judgment may be involved since presumably the similarity of the contesting marks does not rise to the level of infringement but nevertheless might be causing a deleterious effect on a plaintiff's "famous" mark.

DEFINING DILUTION

As noted in Figure 14A.1, one of the critical decisions to be made in FTDA litigation is whether dilution has occurred, and perhaps to what extent and with what motive, though these seem to be secondary questions. As the courts and litigants have wrestled with this question, the most popular criteria seem to be related to an examination of the contesting marks themselves.

Analyzing the Marks

The law itself provides the primary standards for identifying dilution, although that definition is not clear to all. Others have taken their turn. The FTDA standards were summarized by Shire, who noted that in order to prove dilution, a plaintiff must show that:

- The mark is famous
- The defendant is using the mark commercially
- The defendant's use began after the mark became famous
- The defendant's use of the mark *dilutes the quality of the mark by diminishing the capacity of the mark to identify and distinguish goods and services.*[6]

[4] 15 U.S.C. 1114, (1)(a) and (b).
[5] 15 U.S.C. 1125, (c)(1).
[6] *Panavision International L.P. v. Toeppen*, 141 F3d 1316, 1324, 46 USPQ2d 1511, 1518 (CA 9 1998), emphasis added. As referred to in Howard J. Shire "Varying Standards for Assessing Whether There Is Dilution Under the Federal Trademark Dilution Act," *TMR* 91, No. 6 (November-December 2001).

The courts have had their opportunity to describe dilution as well. As reported by Shire,[7] one of the first opinions addressing dilution (though predating the FTDA) was in *Mead Data Central Inc. v. Toyota.*[8] Here six factors were enumerated, two of which focus on the marks themselves:

1. Similarity of the marks
2. Similarity of the products

In *Nabisco*[9] and *Federal Express*,[10] the Second Circuit listed 10 factors, of which five are directed toward an examination of the subject marks:

1. Distinctiveness of senior user's mark
2. Similarity of the marks
3. Proximity of the products and the likelihood that the senior user will bridge the gap
4. Interrelationship among the distinctiveness of the senior mark, the similarity of the marks, and the proximity of the products
5. Adjectival or referential quality of the defendant's use

In *Eli Lilly & Co.*,[11] *I.P. Lund Trading ApS*,[12] *Hasbro, Inc.*,[13] and *Luigino's Inc.*,[14] the First, Seventh and Eighth Circuits took a less complex view and emphasized "similarity of the marks" in their analyses.

Defining dilution by analyzing the involved marks or the products/services they represent has proven unsatisfying in most cases. Some have recognized the concept of "tarnishment" as a form of dilution. Hartman described it:

> Tarnishment occurs when a famous mark is used on products of shoddy quality, or is portrayed in an unwholesome or unsavory context—usually sexual, obscene or illegal activity—likely to evoke unflattering thoughts about the owner's product . . . [or more

[7] Shire, "Varying Standards," fn 6.
[8] 875 F2d 1026, 1032–40, 10 USPQ2d 1961, 1966–73 (CA 2 1989).
[9] *Nabisco, Inc. v. PF Brands, Inc.*, 191 F3d 208, 51 USPQ2d 1882 (CA 2 1999) as reported by Shire.
[10] *Federal Express Corp. v. Federal Espresso, Inc.*, 201 F3d 168, 176-77, 53 USPQ2d 1345, 1351-52 (CA 2 2000) as reported by Shire.
[11] *Eli Lilly & Co. v. Natural Answers, Inc.*, 233 F3d 456, 468–69, 56 USPQ2d 1942, 1951 (CA 7 2000).
[12] *I.P. Lund Trading ApS v. Kohler Co.*, 163 F3d 27, 49 USPQ2d 1225 (CA 1 1998).
[13] *Hasbro, Inc. v. Clue Computing, Inc.*, 232 F3d 1, 56 USPQ2d 1766 (CA 1 2000).
[14] *Luigino's Inc. v. Stouffer Corp.*, 170 F3d 827, 50 USPQ2d 1047 (CA 8 1999).

> broadly] . . . associations contrary or antithetical to those engendered by the senior mark."[15]

Analysis of the involved marks or products/services can be useful in this type of situation. If some form of tarnishment is not evident, however, then an analysis of the contesting marks themselves is unlikely to be helpful.

Focus on the Consumers

In a number of cases, the courts have focused on the characteristics of consumers or the reactions of consumers (from survey evidence) to the alleged dilutive activity. A number of criteria were cited (case references footnoted previously):

- Sophistication of consumers (*Mead*)
- Shared consumers and geographic limitations (*Nabisco* and *Federal Express*)
- Sophistication of consumers (*Nabisco* and *Federal Express*)
- Actual confusion (*Nabisco* and *Federal Express*)
- At a minimum, consumers will possibly identify the two products with the same mark (*Hasbro*)
- Whether consumers connected the famous mark with the defendant's products (*Luigino's*)

When the focus is on consumer characteristics, it would seem that an attempt is being made to detect the extent to which the involved consumers *ought* to be affected by the activity. Going the next step—presenting survey evidence—is presumably aimed at detecting to what extent a type of dilution called "blurring" has taken place.

When something is "distinctive" (a quality that the FTDA attributes to a famous mark), there exist some bright lines around it that clearly differentiate it. When those lines are permitted to blur, some distinctiveness is lost. Pokotilow and Fefferman discuss blurring:

> While the specific harm caused by dilution by blurring can be difficult to quantify, the legislative history of the FTDA clarifies that,

[15] Steve Hartman, "Brand Equity Impairment—The Meaning of Dilution," *TMR* 87, No. 4 (July-August 1997): 424–425.

> over time, dilution "of a famous mark reduces the public's perception that the mark signifies something unique, singular, or particular."[17] The harm caused by dilution occurs gradually, and therefore, is not necessarily amenable to measurement at a particular time.[16]

Their article provides a very good overview of survey evidence relative to dilution issues and concludes that a carefully done survey can assist in establishing the presence or absence of dilution by blurring, even though "survey evidence offered for proof of fame and dilution under the FTDA will likely be scrutinized by the courts with even greater suspicion than surveys presented in source confusion cases under the Lanham Act."[18]

Just as with an analysis of the contesting marks themselves, attempting to discover what consumers think also has, in the main, been an unsatisfactory way to define or detect dilution. One reason for this may be that vigilant trademark owners act promptly toward possible encroachments on their trademark rights. There may not be sufficient time for consumers to become aware of what might be, if left unattended, a source of confusion.

Enter Damages

Perhaps since the evidence provided from an analysis of the marks, products and services, and consumer feelings often falls short of being conclusive, the concept of economic damage entered the courts' thinking. The Lanham Act provides for monetary relief, in the case of infringement, in the form of "any damages sustained by the plaintiff."[19] While primarily a measure of monetary relief, plaintiff damages also can form part of the proof of actual harm caused by an infringement. It is a short leap of logic, then, to consider the use of plaintiff damages as a proof of dilution as well, even though monetary relief may not be the result.

The opinion in *Ringling Bros.*[20] brings together the concepts of analyzing the marks and products/services as well as looking to consumers for guidance in detecting dilution. It also introduces the concept of economic damage as a "litmus test" for dilution. In this case, Ringling Bros.-Barnum & Bailey Combined Shows, Inc. argued that the use of the phrase "The Greatest Snow on

[16] HR Rep. No. 104-374 at 3 as cited by Steven B. Pokotilow and Stephen A. Fefferman, "FTDA Survey Evidence: Does Existing Case Law Provide Any Guidance for Conducting a Survey?" *TMR* 91, No. 6 (November–December 2001).
[17] Potokilow and Fefferman, "FTDA Survey Evidence," pp. 1152–1153.
[18] Ibid.
[19] 15 U.S.C. 1117(a).
[20] *Ringling Bros.-Barnum & Bailey Combined Shows, Inc. v. Utah Division of Travel Development*, 170 F3d 449, 458, 50 USPQ2d 1065, 1072 (CA 4 1999).

Earth" by the Utah Division of Travel Development was dilutive. In its opinion, the court cited factors relating to the marks themselves, factors relating to consumers, and evidence of economic damage. It introduced a standard requiring a plaintiff to prove:

- A sufficient similarity between the junior and senior marks to evoke an "instinctive mental association" of the two by a relevant universe of consumers, which is the effective cause of an actual lessening of the senior mark's selling power (i.e., its capacity to identify and distinguish goods or services)
- The court cited three ways to prove "actual lessening":
  1. Proof of an actual loss of revenues
  2. Consumer survey evidence demonstrating "mental association" of the marks and further consumer impressions from which actual harm and cause might rationally be inferred
  3. Relevant "contextual factors," such as the extent of the junior mark's exposure, the similarity of the marks, the firmness of the senior mark's hold.

The court here not only introduced the economic damage concept but refined it to the extent that the economic damage presented must show *actual, currently evident* damage (in the form of an actual loss of revenues) instead of only the likelihood of future damage. The plaintiff (understandably) was not able to prove an actual loss of revenues. The court's resistance to considering "likely" and "future" events may stem from a traditional suspicion of "speculative" evidence, but this, in our view, is a critical issue.

The Ringling opinion was one that brought into focus what remains as a fundamental question regarding the enforcement of the FTDA: Must a plaintiff prove actual, current injury, or is proof of the likelihood of injury sufficient?

The U.S. Supreme Court soon will consider these questions and the interpretation of the FTDA as it relates to the form of proof that a plaintiff must present to receive remedy in a trademark dilution case. The case in point stems from *V. Secret Catalogue, Inc. v. Moseley*,[21]. V. Secret Catalogue, Inc. ("VSC") sells women's lingerie and apparel under the "Victoria's Secret" trademark and brought suit against Victor and Cathy Moseley, who operated a store offering lingerie and adult novelties under the name "Victor's Little Secret" (formerly "Victor's Secret"). The federal district court for the Western District of Kentucky rendered judgment in favor of VSC on the issue of dilution. It found that there

[21] 259 F. 3d 464, 475-476 (6th Cir. 2001).

had been a "blurring" of the distinctiveness of the VSC mark and that "tarnishment" had occurred as well, due to the risqué nature of the Moseley business. The U.S. Court of Appeals for the Sixth Circuit affirmed this decision.

Because of the considerable disparity among the circuit courts relative to the proof required in a dilution matter, the Moseleys petitioned the Supreme Court on the question of whether the owner of a famous mark must prove actual, current injury, or whether proof of the likelihood of injury is sufficient. There rests the question at this writing.

In our opinion, the courts that have focused on economic damages as a tool to detect dilution have been on the right track, but those that have required proof of actual, current economic damages while barring consideration of future events have misunderstood some critical elements of damage theory.

ECONOMIC DAMAGES AND DILUTION

Basic Damage Theory

The basic theory of damages suggests that the:

> Market value of a business or property prior to an alleged damaging event
>
> *Less:* Market value of the same business or property after the alleged damaging event
>
> *Equals:* Economic damage to the business or property due to the event

It is useful to examine the components of this equation. *Market value* has been variously been described as: expressible in terms of a single lump sum of money considered as payable or expended at a particular point in time in exchange for property, that is, the right to receive future benefits beginning at that particular timepoint[22] and as " the amount at which a property would exchange between a willing buyer and seller, neither being under compulsion, each having reasonable knowledge of relevant facts, and with equity to both." Market value also has been described as the present value of the future economic benefits of ownership.

An appraisal is a process by which we put an imaginary buyer and seller together and attempt to describe the results of a transaction they might negotiate. An appraisal represents a transaction that has not taken place, nor may it ever. But it is useful to think in terms of a buyer considering the purchase of an asset

[22] Henry A. Babcock, FASA, *Appraisal Principles and Procedures* (Washington, DC: American Society of Appraisers), ch. 6, p. 95.

undamaged as compared with the same buyer considering a purchase after a damaging event.

The concept of *present value* is based on the premise that a dollar to be received in the future is worth less than a dollar in one's pocket today. For our purpose, however, the central fact is that value is always dependent on the *expectations* of buyers and sellers, or their *perceptions of future events*: "The cost of capital represents investors' *expectations* . . . [and] is applied to *expected economic income*."[23]

Essential to any valuation is a clear definition of the property appraised. Property has been described in terms of a "bundle of rights":

> The bundle of rights theory refers to the concept that ownership of real property is embodied in a number of separate privileges. These include the right to occupy it and use it; the right to sell, merge, donate, mortgage or bequeath it; and the right to transfer by contract some of the benefits for a period of time."[24]

Included also is the right *not* to do any of these things. This is the important element of control. A buyer of property rights is certainly going to be sensitive as to which of these rights are included in the transaction. The price that a buyer is willing to pay for property rights is directly related to the extent of rights received. Therefore, any measurement of value (acting as a surrogate for an actual transaction) must consider which of these rights are present and which are not.

With these principles in mind, we can begin to describe the characteristics of a trademark "before the alleged event" in the damage equation.

Trademarks in a Perfect World

The FTDA is intended to protect trademark owners from harm to their property rights. The "harm" is, however, not well defined and subject to varying interpretations. As a precursor to discussing our views on harm, it is useful to observe trademark rights *unharmed*. The following paragraphs are taken from Smith's book *Trademark Valuation*.[25]

> A trademark is used to identify the source of a product or service and to distinguish that product or service from those coming from other sources. As defined in the Trademark Act of 1946 (the Lanham Act), a trademark is "any word, name, symbol or device or any combina-

[23] Shannon P. Pratt, *Cost of Capital* (New York: John Wiley & Sons, Inc. 1998), p. 5.
[24] Gordon V. Smith, *Trademark Valuation*, (New York: John Wiley & Sons, Inc., 1999), p. 160.
[25] Ibid.

tion thereof [used by someone to] identify and distinguish his goods, including a unique product, from those manufactured or sold by others and to indicate the source of the goods." A trademark also serves as an assurance of quality—the consumer comes to associate a level of quality with the goods or services bearing a given trademark. Trademarks have been described as the embodiment of goodwill. The courts have addressed these aspects of trademarks in various ways:

> Trademarks help consumers to select goods. By identifying the source of the goods, they convey valuable information to consumers at lower costs. Easily identified trademarks reduce the costs consumers incur in searching for what they desire, and the lower costs of search the more competitive the market."[26]
>
> A trademark also may induce the supplier of goods to make higher quality products and to adhere to a consistent level of quality. The trademark is a valuable asset, part of the "goodwill" of the business. If the seller provides an inconsistent level of quality, or reduces quality below what consumers expect from earlier experience, that reduces the value of the trademark. The value of a trademark is in a sense a "hostage" of consumers; if the seller disappoints the consumers, they respond by devaluing the trademark.[27]
>
> The protection of trade-marks is the law's recognition of the psychological function of symbols. If it is true that we live by symbols, it is no less true that we purchase goods by them. A trademark is a merchandising short-cut which induces a purchaser to select what he wants, of what he has been led to believe he wants. The owner of a mark exploits this human propensity by making every effort to impregnate the atmosphere of the market with the drawing power of a congenial symbol . . . to convey, through the mark, in the minds of potential customers, the desirability of the commodity upon which it appears. Once this is attained, the trade-mark owner has something of value."[28]

Other writers have differentiated the brand from the product by noting that a product is what is manufactured for sale, while a brand is what the customer buys. Kapferer presents the brand in this way:

[26] *Scandia Down Corp. v. Euroquilt, Inc.*, 772 F2d 1423,1429 (7th Cir., 1985), *cert. denied*, 475 U.S. 1147 (1986).
[27] Ibid.
[28] *Mishawaka Mfg. Co. v. Kresge Co.*, 316 U.S. 203, 205 (1942).

> For the potential customer, a brand is a landmark. Like money, it facilitates trade. . . . One word, one symbol summarizes an idea, a sentence, and a long list of attributes, values, and principles infused into the product or service. A brand encapsulates identity, origin, specificity, and difference. It evokes this information-concentrate in a word or a sign.
>
> Like money once again, brands facilitate international trade. Brands are the only true international language—a business Esperanto.[29]

In his discussion of the role of brands, Aaker relates:

> A brand is a distinguishing name and/or symbol (such as a logo, trademark, or package design) intended to identify the goods or services of either one seller or a group of sellers, and to differentiate those goods or services from those of competitors. A brand thus signals to the customer the source of the product, and protects both the customer and the producer from competitors who would attempt to provide products that appear to be identical."[30]

In his book *Brand Asset Management*, Davis describes how strong brands benefit their owners:

1. Loyalty drives repeat business. It is estimated that the lifetime value of a customer of one of P&G's brands is several thousand dollars.
2. Brand-based price premiums allow for higher margins.
3. Strong brands lend immediate credibility to new product introductions.
4. Strong brands allow for greater shareholder and stakeholder returns.
5. Strong brands embody a clear, valued, and sustainable point of differentiation relative to the competition.
6. Strong brands mandate clarity in internal focus and brand execution.
7. The more loyal the customer base and the stronger the brand, the more likely customers will be forgiving if a company makes a mistake.

[29] Jean-Noel Kapferer, *Strategic Brand Management*, (London: Kogan Page Limited, 1992), p.10.
[30] David A. Aaker, *Managing Brand Equity: Capitalizing on the Value of a Brand Name* (New York: The Free Press, 1991), p. 7.

8. Brand strength is a lever for attracting the best employees and keeping satisfied employees.
9. 70 percent of customers want to use a brand to guide their purchase.[31]

The value of a trademark is the present value of the future economic benefits that it can produce for its owner. Typically, we value a trademark under the assumption that a buyer will acquire all of the rights in the "bundle." The buyer, then, can expect to garner the economic benefit from *all reasonable exploitations*.

If, however, the bundle of trademark rights has been diminished by the owner's actions or otherwise, the buyer's opinion of value will be diminished as well because the buyer's *expectation* is that future economic benefits will be less than otherwise.[32] If, at any time, a trademark owner cannot deliver the full bundle of rights to a prospective purchaser, then the trademark's value is less than it would otherwise be. This is true whether the owner has given, sold, or licensed away the rights, or whether there is simply the *threat of loss* (i.e., infringement or dilution litigation under way) or if there has been some uncompensated *erosion of rights* (i.e., unrecognized dilution).

Trademarks in the Real World

In the real world, trademarks are subject to continual assault by competition, by events, by technology, and even by their owners. One can never blindly assume that a trademark owner always controls the complete bundle of rights.

Self-inflicted Wounds

Even trademark owners themselves can inadvertently diminish the rights:

> In the 1930's, tennis player René LaCoste began a line of cotton sportswear bearing his name—and an alligator trademark. General Mills acquired the licensing rights in the 70's and garnered millions in its exploitation, stretching the brand to, and perhaps beyond, its limits. So successful was this that "knock-offs" proliferated as well. The result of all this was that the brand was severely degraded. General Mills sold the rights to its Crystal Brands group, purchased by management. Their efforts on its behalf were disappointing, and the

[31] Scott Davis, *Brand Asset Management* (San Francisco: Jossey-Bass, 2000).
[32] For a more extensive discussion of this, see Smith, *Trademark Valuation*, Chapter 7, "Trademark Royalty Rates."

> rights were sold to Devanlay, S.A., a French company with a long association with the brand. Devanlay took the brand off the market for awhile and in 1994 brought it back on a premium priced shirt like the original. Early reception was good, and time will tell whether this rejuvenation will be successful.[33]

Davis cites other examples of how brand extensions can go wrong.[34] Merrill Lynch in 1999 decided to enter the online brokerage business. It then became concerned that this move might denigrate its traditional "full-service" image and put it on a parity (in consumers' minds) with "bargain" brokers. An advertising campaign ensued to reassure its customers that Merrill Lynch would "still be there" for them.

In discussing the dangers of brand extensions, Davis quotes Brad VanAuken, former director of brand management at Hallmark: "You have to be careful how far you extend the brand. You have to be very careful in clarifying what you stand for. At Hallmark, we were always riding that fine line of how far we can go with our positioning of caring shared, without losing its inherent meaning."

In his article, "Dilution Redefined for the Year 2002," Jerre B. Swann Sr. cites many instances of unfortunate brand extensions and discusses how these dissonances affect the meaning of the core brand and reduce its economic effectiveness.[35] It is easy to relate these events to the dilution schema.

Of course, these "self-inflicted wounds" are not the subject of litigation. But one can easily draw the analogy between these events and a dilutive act by a third party that could be thought of as an unauthorized brand extension.

Degrading Brand Equity

Hartman describes this diminution of rights in terms of its effect on brand equity:

> Brand equity allows its owner to define an identity or persona for itself and its products; differentiate itself and its products from those of its competitors; preempt competitive efforts; enhance brand awareness; create a preference for its products; reduce market entry costs for new products; and enable its products to command a premium price.
>
> As I see it, dilution is the impairment of brand equity caused by a use of the mark that creates associations and images inconsis-

[33] Ibid., p.114.
[34] Davis, *Brand Asset Management*, pp.139–143.
[35] Jerre B. Swann Sr., "Dilution Redefined for the Year 2002," *TMR* 92, No. 3 (May–June, 2002).

> tent with the equity. Dilution erodes or (in the words of the statute) "lessens" the brand's identity, and in so doing, weakens the brand's pricing power, ability to command market share, and durability, those qualities of brand equity that account for its selling power and marketing value.
>
> ". . . . in contrast to infringement, a trademark owner can dilute his own mark. The use of a mark in a manner that impairs equity is dilutive regardless of who is responsible for the impairing use.[36]

Hartman's last point is interesting (and correct) in that it helps to define what dilution is and is not. As an example, it is said that the value of the PIERRE CARDIN mark has been degraded by its unduly wide use on a wide range of products. If true, this is a case in point and an example of dilution in its classic sense: "to lessen the potency, strength, purity, or brilliance of by admixture."[37]

A Stock Market Analogy

The "dilution" terminology is also used in another context. Stockholders in corporations are very sensitive to actions that may dilute the value of their proportionate ownership. Assume, as an example, that a corporation has 1 million shares of stock issued and that these shares are owned by 10 shareholders, each with 100,000 shares. Each shareholder owns 10% of the corporation. If the corporation issues 100,000 new shares to a new investor, the old shareholders interest in the corporation is *diluted* to 9.1%. If there was no change in the corporation except the issuance of the new shares, the *value* of the old shareholders interest went down by 10%.

Common stock values are based, in large part, on an investor's expectations about future earnings and dividends. When those future earnings and dividends must be shared by a larger number of shareholders due to dilution, then the value goes down. This echoes the Hartman's comments on brand equity.

Brand Exploitation

In discussing ways to leverage a brand, Aaker presents several scenarios:

- Line extensions in the existing product class
- Stretching the brand vertically in the existing product class

[36] Hartman, "Brand Equity Impairment," pp. 420–422.
[37] American Heritage Dictionary (Boston: Houghton Mifflin Company, 1985).

- Brand extensions in different product classes
- Cobranding[38]

Aaker touches on the risks associated with these strategies. His comments on "stretching a brand vertically" seem particularly relevant: "brands move down easily (if sometimes inadvertently), and they find that there are problems and challenges created by getting to the bottom. The biggest challenge is to avoid harming the brand, particularly in terms of its perceived quality associations." It seems to us that a dilution situation is quite representative of a downward "stretch" of a trademark. The law even speaks of "senior" and "junior" marks. It seems very likely that consumers would see the presence of a junior mark as a downward movement in the senior mark, and as Aaker notes, "The problem is that moving down affects perceptions of the brand perhaps more significantly than any other brand management option. Psychologists have documented the fact that people are influenced much more by unfavorable information than by favorable information." Thus we may infer that a consumer, confused by a dilutive act, may be more negatively affected by dilution than otherwise.

Brand managers strive to maintain the clear image of their brands' "personality" while at the same time exploiting them to increase sales and profits. Aaker notes that an unsuccessful brand extension is unlikely to harm the original brand if the original brand associations are very strong and if there is a material difference between the products or services. At the same time, he notes that "Cadbury's association with fine chocolates and candy certainly weakened when it got into such mainstream food products as mashed potatoes, dried milk, soups, and beverages."[39]

While ill-conceived brand extensions are a form of "self-dilution" (reducing value), it is equally clear that successful brand extensions can be an extremely lucrative form of trademark exploitation and can add greatly to value.

Loss of Control

The Harley-Davidson success story is just that because its trademark was once in serious jeopardy due to the inattention of its once owner and a poor response to overseas competition: As Timothy K. Hoelter describes it, 'There was a strong risk back then that the Harley-Davidson trademark would go the way of Escalator and Aspirin, because it was being used by unauthorized people with abandon."[40]

[38] David A. Aaker, *Building Strong Brands* (New York: The Free Press, 1996), p. 275ff.
[39] David A. Aaker, *Managing Brand Equity*, pp. 222–223.
[40] Smith, *Trademark Valuation*, p. 118

The Harley-Davidson problems included authorized dealers who had allowed their facilities to deteriorate as well as unauthorized motorcycle shops that advertised themselves as dealers. This, together with the unauthorized use of the H-D trademarks on shoddy merchandise, was tarnishing the Harley-Davidson image. In addition, there were those who applied the H-D trademarks to goods that, while they may have been of good quality, represented "brand extensions" of which the company was unaware. The trademark misuse ranged from counterfeiting to infringement and probably included activities that would have come within the dilution purview. In short, the company had lost control of its marks, and it worked hard and very successfully to regain it.

From this troubled time to the present, the value of the Harley-Davidson company has grown tremendously. It is obvious that the value of its trademark portfolio has grown as well. This change in value is clearly a measure of the effects of the deleterious trademark activities. This example illustrates the possible effect of losing control of one's trademark rights, because in losing control one does not know which or how many of the bundle of rights are in jeopardy.

Trademark Portfolios

Many trademark portfolios are pyramidal with well-defined levels. Aaker gives an example:

Corporate Brand	Nestlé
Range Brand	Carnation
Product Line Brand	Carnation Instant Breakfast
Subbrand	Carnation Instant Breakfast Swiss Chocolate
Branded Feature/ Component/Service	NutraSweet[41]

When the associations between trademark levels are strong, harm to one may bring harm to others, or even all.

> As an example, let us observe a small manageable enterprise—General Motors. Its brands can be thought of as a pyramid. General Motors, or GM, is the trademark at the top of the pyramid. It is the "endorsing brand"[42] of a large family of trademarks encompassing not only those for automobiles, but also those for trucks, financial services, parts, and

[41] Aaker, *Building Strong Brands*, p. 242.
[42] Kapferer, *Strategic Brand Management*, p. 118.

a whole host of other manufactured products. The Buick brand is one of the automobile brands. It, in turn, is an umbrella brand for several automobile models, such as Riviera, Park Avenue, and Century. We can build this pyramid almost endlessly, for there are accessories, car care products, ball caps, golf shirts, replicas of old dealer signs, and toy vehicles all bearing some trademark or other in the General Motors family. Some of these brands are being exploited by GM, others by licensees. The exploitation of these many brands has an effect on the others, to some greater or lesser degree. The influence can be negative or positive (or neutral).

A valuation task may seem simpler if we start at either the top or bottom of the pyramid. We could begin with the value of General Motors as a whole and from there estimate the value of the whole body of GM trademarks. In that value we would have captured all of the good things (from an economic standpoint) and all of the bad things. We would have reflected all of the interrelated effects.

At the bottom, we could analyze the income-producing capability of the ball cap business, which would probably be that of a licensee. That would be a straightforward process, focused on the stream of royalty income to GM. In the center section of the pyramid, the small income streams are becoming creeks are becoming small rivers are becoming tributaries are becoming rivers of income flowing into the General Motors lake. When we measure the income (and therefore the value) at any point, we are quantifying all the positive and negative effects below it. As an example, suppose the ball cap licensee is paying royalties that contribute to the overall earnings of the corporation. Suppose, however, one Buick ball cap purchaser was unhappy with its quality and, because of that, purchased a Pontiac instead of a Buick. When we valued the Buick brand, we would (theoretically) observe the effect on earnings of that one unsold automobile, and the effect would be felt in the value of the Buick brand. So, while we might not be aware of this specific occurrence, we will catch up with its effect somewhere in the chain of earnings.[43]

Lost Opportunity

A very important form of economic harm is a lost opportunity. In the business world, lost opportunity can manifest itself in several forms. The failure to obtain a contract from a supplier in a time of rising prices is an example. A successful

[43] Smith, *Trademark Valuation*, pp. 67–68.

encroachment in one's market territory by a competitor can be another. The latter situation can arise with a trademark.

In September 1953 General Motors Corporation received trademark registration for its then nascent but now well-known Corvette series of automobiles. As the popularity of the Corvette sports car grew, General Motors registered the mark for:

Toys and playthings (1987)

Jewelry (1988)

Eyeglasses (1988)

Guitars (1996)

Coin and non–coin-operated pinball machines (1996)

Cologne (1997)

Cookies (1998)

Clothing, footwear, and headgear (2000)

Metal money clips, metal gear shift knobs, tin signs, key rings (2001)

General Motors does not manufacture jewelry or cookies or cologne, so it is logical to assume that these registrations were for the purpose of exploiting the well-known Corvette mark through licensing. It is interesting to consider what might have happened if a third party had attempted to register a Corvette mark for jewelry, eyeglasses, guitars, or cologne and if the FTDA had been in effect at the time. Presumably if the third-party use had not copied GM's logo, type font, unique colors, or directly alluded to the automobile, the new use might not have risen to the level of infringement. Dilution might well have been the issue between GM and the third party.

It is extremely unlikely that GM could have shown that sales of its Corvette automobiles had suffered as a result of the use of the mark by the Corvette Guitar Company.[44] It also may have been unable to show rampant confusion among guitar purchasers. Consideration also might have been given to the fact that there is considerable "market distance" between sports cars and guitars, and that sports car and guitar purchasers are "sophisticated" enough not to be misled by the name similarity. As a result, GM might well have lost its dilution claim.

The critical point, however, is that GM also would have lost the opportu-

[44] We believe that GM's license was with the Gibson Guitar Company, but we have not used that firm's name herein.

nity to license its well-known mark for use on a guitar, a potential income stream that we now know was viable. The inescapable conclusion is that a dilutive activity can preclude exploitation opportunities and therefore reduce value, resulting in some measure of actual damage. As we pointed out previously, a prospective buyer of GM's Corvette mark would recognize the reduced right to license the mark in a lower than otherwise purchase price.

The *Ringling Bros.* case cited previously could be thought of as exemplifying economic harm from a lost opportunity. It is not inconceivable that the State of Utah might have approached Ringling Bros. to negotiate a license.

Lost trademark exploitation opportunities can vary greatly, depending on the versatility of the mark. Some marks are narrow in scope, and the opportunities to exploit them by brand extension are likewise few. The loss of one of these opportunities may be more serious than when the possibilities for trademark exploitation are broad.

Summary

This discussion has pointed out the types of deleterious events that can occur in the life of a trademark. These events either cause one to lose some of the bundle of rights or impair one's the ability to exploit the full bundle of rights:

- Lost exploitation opportunities
- Unsuccessful brand extensions
- Reduced effectiveness in the core use of the mark, degradation of brand equity
- Loss of control

In our opinion, these situations can have measurable economic impact. They also are situations that can be caused by activities described as dilution. We are therefore describing the alleged "event" in the damages equation.

We suggest that a trademark collision that has no economic impact on the trademark owner does not require remedial action. In basketball, this would be called a "no harm, no foul" situation. Applying the damage equation, value before the alleged event equals value after the alleged event, so damage is zero.

It seems reasonable to assume that legislators, in framing trademark laws, sought to put in place a mechanism by which those who suffer *economic* harm can seek remediation. Therefore, proof of economic harm should be the critical test. But "economic harm" should be measured using valuation and investment theory, and should not be narrowly limited (e.g., "actual, current lost profits") to make the test easier to apply.

BACK TO VALUATION

In our opinion, the proper test for economic harm involves a comparison of market value as suggested by the damage equation referred to previously. This is not an entirely new concept.

An Accounting Analogy

Typically, intangible assets such as trademarks are not reflected on a company's balance sheet at anything approaching their market value. There are circumstances, however, when this does not apply and trademarks appear on a company's balance sheet. Since trademarks are long-lived assets, there is then a concern that the amount carried on the books of the company is reflective of the value of the trademarks in the ongoing business. As an example, if a trademark is abandoned, its carrying amount must be removed from the books. This is an easily defined event with an understandable result.

In a more difficult task, companies also must regularly evaluate whether long-lived and nonamortizable assets such as trademarks are *impaired* in value, even though they may still be in use. Accounting rules suggest that events triggering an evaluation might include:

- A significant adverse change in legal factors
- A significant adverse change in the business climate
- An adverse action or assessment by a regulator
- Unanticipated competition

To quantify impairment, a company must estimate the future cash flows attributable to the trademark (i.e., cash flows that will be realized from its future use in the business). If the trademark's market value or the present value of future cash flows is less than the carrying amount, an impairment loss is recognized and the carrying amount is adjusted to that calculated fair value.[45] When such an adjustment is made, the company must disclose in its financial statements:

- A description of the impaired intangible asset and the facts and circumstances of its impairment
- The amount of the impairment loss and the method used to estimate it

[45] Further information is contained in Financial Accounting Standards Board (FASB) Statement of Financial Accounting Standards No. 142, "Goodwill and Other Intangible Assets." See also 2001 Supplement Chapter 4A.

- The location in the income statement in which the loss is reported
- The business segment in which the asset is reported

We are witnessing a considerable amount of impairment at the present time, as the e-commerce bubble deflates and as the financial impact of the telecommunications overcapacity is felt. Many long-lived intangible assets, including trademarks, are judged to have become impaired, and significant balance sheet write-downs have been in the business news.

One can easily understand how trademark issues such as infringement and dilution could trigger an impairment evaluation. A trademark that has been judged to be infringing would likely be the subject of an impairment analysis. A trademark that a plaintiff has alleged to have been diluted by the actions of another also might qualify for an impairment evaluation. This is true especially if a court has declined the dilution argument and not taken injunctive action. A plaintiff might be bound by accounting rules to analyze the effect of dilution on future cash flows and report an impairment even if a court finds no dilution.

This method for quantifying impairment represents a close analogy to what we advocate as a method for detecting trademark dilution. An analysis of market value is the key. Thus Figure 14A.1 can be revised to Figure 14A.2.

Market Value Analysis

To perform the market value analysis that we are suggesting to take the place of other tests for dilution, one should:

- Analyze the enterprises of the protagonists—specifically with regard to the businesses associated with the trademarks in question—and consider:

    —Markets served

    —Customer characteristics

    —Customer opinions

    —Likelihood of continued convergence

    —Reasonable exploitation possibilities

    —Risks associated with exploitations

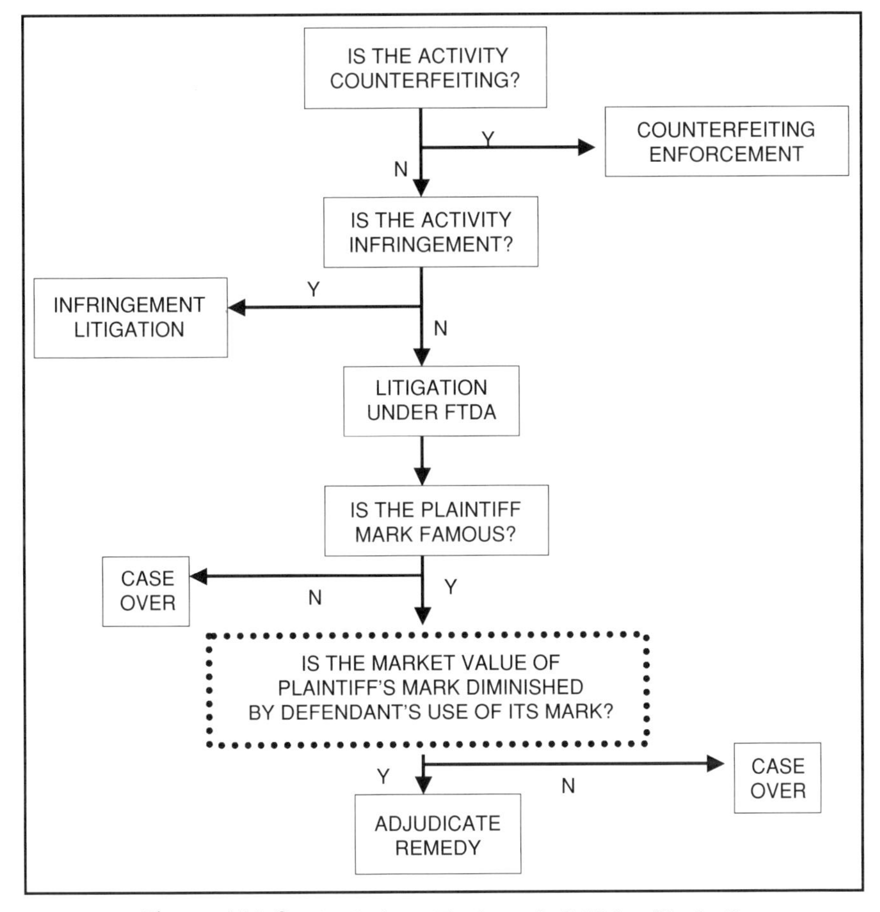

Figure 14A.2 **Analyzing a Trademark Collision (Revised)**

—Reactions from competitors

—Versatility of the marks

- This analysis may require a number of professional disciplines:

  —Econometrics

  —Market research

  —Valuation

- Identify the probable events that may result from the defendant's continued use of its mark.
- Estimate the future economic impact of these events on the plaintiff's mark.
- Calculate the present value of the future income attributable to the plaintiff's mark, assuming no use by the defendant of its mark.
- Calculate the present value of the future income attributable to the plaintiff's mark, assuming continued use by the defendant of its mark.
- Consider other valuation methodologies (direct market or cost approach if appropriate).
- Subtract affected market value from unaffected market value to identify the presence of economic harm from defendant's continued use of its mark.

SUMMARY

One can easily observe from the elements just presented that if the principle of economic harm is used to detect dilution, then:

- One needs to focus on the market value of the plaintiff's mark.
- One needs to focus on the impact on that value of the defendant's continued use of its mark.
- One therefore must be willing to consider evidence concerning future events because value is always "forward-looking." Our entire system of business, investing, and banking relies on estimates of future conditions and events. There is no reason to summarily dismiss such evidence when it relates to trademark dilution.
- One cannot limit consideration to "current, actual lost profits," because while the *presence* of this condition may indicate economic harm, the *absence* of "current, actual lost profits" does not mean that the plaintiff has suffered no economic harm. Again, quick action by the plaintiff may forestall lost profits, or there may be enough market distance between the plaintiff and the defendant that there may never be lost profits in the plaintiff's core business.
- The distinction between "current harm" and "future harm" is meaningless in terms of valuation and investment theory. Future harm *is* current

harm. When we quantify market value, "the future is now." We must consider "likelihood," "possibility," "future" because that is the way the economic world works.

Are these principles sound? We are convinced of it. Is this analysis easy to do? Not at all. Is this analysis expensive? Very likely.

But the alternative is to continue to attempt to work through what may be a murky semantic analysis (are the marks alike?), or to attempt a definitive survey of customer attitude (not an easy or inexpensive task), or to apply an economic damage metric (show current lost profits) that is simply wrong.

17A (New)

The Valuation of Naming Rights

INTRODUCTION

Most sports fans, and many others for that matter, would instantly recognize Lambeau Field as the home of the Green Bay Packers football team. Built in 1957, the stadium was renamed Lambeau Field in 1965 following the death of E.L. "Curly" Lambeau, the founder and first coach of the Packers. Lambeau Field is to undergo a major renovation and modernization. This sort of activity often triggers consideration of the "naming rights" issue. That is, should the stadium be renamed and bear the corporate logo of some enterprise that would, in exchange, be willing to pay for the privilege? In this particular case, that decision is up to the city fathers and citizens of Green Bay, Wisconsin, who own the team.

While there has been understandable consternation over the possible renaming of this venerable facility, the Green Bay City Council approved a proposal that will permit the team and Brown County to pursue naming rights proposals as early as 2001. The city and the Packers would equally divide the naming rights revenue. On November 7, 2000, voters supported the sale of naming rights by a margin of 53 percent to 47 percent in an advisory referendum. Lambeau Field may then join the many sports facilities that are now adorned with corporate names. This trend, often triggered by the building of a new stadium or the rebuilding of an old one, is gathering momentum as the demand by teams for new facilities puts additional pressure on dwindling public funds for such projects.

Perhaps this all began, albeit somewhat indirectly, in 1926 when William Wrigley, Jr., the owner of the Chicago Cubs baseball team, renamed Cubs Park as Wrigley Field. This venerable ballpark remains Wrigley Field to this day. Inevitably, this name is associated in the public consciousness with the Wm. Wrigley, Jr., Company and its famous chewing gum products. In naming the field, Mr. Wrigley may have just wanted to signify his ownership of the field and the Cubs. But who knows? He may have recognized the commercial possibilities as well—that the Wrigley name would be heard by thousands over the airwaves and seen on countless newspaper sports pages nearly every day of every summer in the upper Midwest. Today, of the 30 Major League Baseball parks open or soon to be opened, 10 clearly bear the names of corporations and the newest eight all bear corporate names.

In spite of the commercial possibilities, local residents are not always in favor of corporate naming. Most of the time, they don't have a say in the matter since the sports facility is usually owned by team owners. In Denver, however, Mile High Stadium belongs to the Metropolitan Football Stadium District and a new stadium for the Denver Broncos football team is being built next to the existing Mile High Stadium. The District had estimated that selling the name could be worth between $52 and $89 million. Invesco Funds Group came in with a bid of $120 million, however, and this was accepted by the District, albeit with some fan discontent. Denver's new basketball and hockey arena bears the name Pepsi Center, and that company is said to have paid a record $68 million for those naming rights.[1] For those interested in the economics of this type of transaction, we note that most naming rights transactions include other benefits such as long-term leases for luxury boxes and suites for the naming corporation, and the terms of the deal can cause the price to vary considerably.

Our research readily revealed more than 100 naming rights transactions, with prices ranging from $250,000 to over $200 million. Facilities of all kinds were involved, ranging from major league sports facilities to local stadiums. The terms of agreement range from five years to 99 years, though the longest term in recent contracts is 30 years. A list of these transactions is contained in Appendix A to this chapter.

Motorsports have entered into this area of commercialization as well. Early in 1999 the Charlotte Motor Speedway was renamed Lowe's Motor Speedway in a 10-year naming rights agreement with gross fees of approximately $35 million. In May 2000, race fans were injured when a pedestrian bridge collapsed at the race facility. While there has been no adverse reaction for

[1] Larry Fish, "Stadium Approved, the Fight in Denver Is What to Name It," *The Philadelphia Inquirer*, Sunday, May 28, 2000.

Lowe's, this points up one of the risks of a naming rights deal for the sponsor. There is always the possibility of adverse events, just as there can be with corporate sponsorship of an event or the use of an individual as spokesperson.

Even shopping malls have been the subject of naming rights transactions. The Mills Corporation sold the naming rights to a mall in Gwinnett County, north of Atlanta, to Discover Financial Services. The mall will become known as Discover Mills for a 10-year period. The reported price of this transaction was estimated at $10 million. Drew Sheineman, senior vice president of marketing for Simon Property Group's BrandVentures unit, has commented that their 280 malls and 2.6 billion annual shoppers represent a very attractive audience for corporate sponsors.[2]

Naming transactions have also reached a very local level such as in Jefferson County, Colorado. The county stadium, used for high school sports, will bear the name of the U.S. West Communications Group for 10 years. The reported price of this transaction was $2 million.

Naming rights transactions are being made outside of the United States as well. The list in Appendix A contains several Canadian deals. In Hamilton, New Zealand, the former Rugby Park is to be known as Trust Waikato Stadium in a $1.5 million deal. ANZ Banking Group captured the biggest naming rights deal to date in Australia, paying over $50 million for the sponsorship rights at Sydney's Olympic stadium, with the rights to take effect on January 1, 2001. In 1999, Colonial Limited is said to have paid $25 million for the naming rights to Melbourne's Docklands Stadium.

NOTABLE DEALS

Some more recent transactions are notable for their richness,[3] and it is useful to observe what may be some of the reasons why:

STAPLES Center	Los Angeles	$5.2 million/year
PEPSI Center	Denver	$3.5 million/year
PHILIPS Arena	Atlanta	$9.4 million/year
AMERICAN AIRLINES Center	Dallas	$6.5 million/year
PSINet Stadium	Baltimore	$5.6 million/year

[2] As reported by Mark Albright, "Naming Rights Soon May Extend to Malls," *St. Petersburg Times*, June 1, 1999.

[3] The dollar amounts shown, as in Appendix 17A, are estimates of the worth of total compensation, and are not the result of a present value calculation.

FEDEX Field	Landover, MD	$7.9 million/year
CMGI Field	Foxboro, MA	$7.6 million/year

STAPLES Center—Some say that this 1997 deal set the tone for large naming rights transactions to follow. Two NBA teams and one NHL team call this venue home and it is situated in the country's second largest media market.

PEPSI Center—It is said that PepsiCo was motivated to assist Denver's effort to attract an NHL franchise and to keep the NBA home team, the Denver Nuggets. Another influence was no doubt the exclusive soft drink franchise and in-arena concessions rights for PepsiCo brands such as Taco Bell, Pizza Hut, and Kentucky Fried Chicken and the many Frito-Lay snack food products.

PHILIPS Arena—This $180 million deal is the third richest to date. Philips plans to utilize the facility as a test bed for a wide variety of its electronic products. Included also is an enhanced relationship with Time Warner and Turner Broadcasting, the details of which have not been made public.

AMERICAN AIRLINES Center—Apparently in competition with Southwest Airlines for the naming rights, American Airlines paid $195 million for the rights to this home of the Dallas Mavericks (NBA) and Dallas Stars (NHL). The naming rights are extensive, with pervasive signage and the use of the teams' personnel in advertising and marketing.

PSINet Stadium—In addition to naming rights, PSINet has the right to set up a Baltimore Ravens web site and to provide Internet services to subscribing fans. More recently, PSINet was reported to have retained an investment banking firm to be an advisor in the possible sale of the company. We cannot foretell how that might affect this naming rights deal.

FEDEX Field—The richest deal to date in terms of total commitment ($205 million), this renamed field will be home to a single NFL team, the Washington Redskins.

CMGI Field—Home to the New England Patriots, CMGI hopes that the naming rights to this facility will help it to build a nationally known brand. The Internet conglomerate's chairman, David Wetherell, commented that the signage alone will provide 2.8 billion ad impressions each year. The radio and television coverage of games will, of course, enhance that figure considerably.

NAMING RIGHTS AS PROPERTY

It is perhaps somewhat of a misnomer to use the term "naming rights" as if it was some specific type of property heretofore unknown. In Chapter 12, we discuss the exploitation of intellectual property and introduce the underlying theory of the bundle of rights associated with property. In essence, so-called naming rights are really part of the underlying bundle of rights possessed by the owner of a special event facility, such as a sports stadium or arena. One of the ways that the owner of such a facility can exploit that property is by permitting others to place advertising messages on it or in it.

Advertising messages placed in and around sports facilities go as far back as any of us can remember. Baseball fields, from little league to the majors, are replete with ad-adorned fences. It is only recently, however, that a market has emerged for corporate identification of the whole facility and, together with that, the willingness of facility owners to give up the right to name the facility in return for a fee.

What we are calling naming rights, then, arise out of a contract between a stadium owner and a third-party entity, usually a corporation. The facility owner gives up a portion of the total rights of ownership—the right to name the facility. The corporation receives the right to display its name and have the facility identified with it for a period of time. While a corporation's trademark is involved, this is not a license of the trademark, since the facility owner gets no trademark rights from the transaction. It is more akin to a lease of real estate and represents a new form of exploitation of land and structure. Hence, many value elements of the transaction are similar to those found in other types of real estate deals—location, size of the market, tenant quality (in this case teams and events), attractiveness of the structure, and so forth. There are other considerations as well, making the naming rights lease somewhat unique.

As a result of a naming rights contract, the facility owner receives:

- Compensation for the naming rights in the form of an up-front payment and/or annual payments for some period of time.
- A contractual income stream, which may help finance renovation or new construction expenditures.
- If the owner is a governmental entity, income to offset taxes or to affect a reduction in the amount of bonded debt.

On the other side of the transaction, the corporate entity receives:

- The public relations benefit of being a good corporate citizen—perhaps acting to help retain the "home team" (especially in the case of a government-owned facility).

- The associated benefits of a prominent advertising program on-site, including signage, employee uniforms, programs, and product or service displays.
- The benefits of an advertising program reaching beyond the site, through radio and television coverage of the events held there.
- Luxury seating and parking for the entertainment of its clients and customers, as well as season tickets and the right to conduct tours.
- A wide range of concession rights including pouring rights, operating or licensing food and/or beverage concessions, operation of restaurants or bars.
- The right to exclusively provide services related to the company's business, such as automated teller machines, telecommunications services, or electronic products.
- Advertising and marketing rights connected with the resident sports team(s) members and staff.

As with any contract, we would expect to find economic benefits for both parties to the agreement. As in any valuation, the quantification of those economic benefits is the keystone.

VALUATION OF NAMING RIGHTS

As we noted in the very beginning of this book, it is critical to have a clear understanding of the specific rights being appraised. This is no less true with respect to the property that we are calling naming rights. We need to establish which side of the transaction is our focus. Are we estimating the market value of:

- A potential naming rights transaction? That is, are we estimating the probable price that a contemplated transaction will bring?
- A consummated naming rights contract from the standpoint of the corporate lessee?
- A consummated contract from the point of view of the facility owner/lessor?

When we are estimating the probable price of a transaction, we are weighing the economic benefits to both parties and estimating how they will be divided between them. In valuations such as this, we often go to the marketplace seeking benchmark information. It is analogous to a lease of office space, which is a common occurrence in the real estate marketplace. The lessor gives up the right of occupancy for some period of time in exchange for lease payments. The amount of those lease payments is, in general, dictated by the marketplace because there is usually competitive office space available in a given market and lessees have the choice of alternative leases or the option of constructing a building themselves for their own occupancy.

With respect to naming rights, the marketplace is not nearly so informative, as we will demonstrate in the following section. There are usually only one or two naming rights opportunities of equal stature in a given metropolitan area. Because of the extreme variety of the terms associated with a naming rights agreement, the value of a naming rights transaction in Dallas a year ago may not be very informative as to an appropriate market value for a naming rights agreement in New York currently.

For the facility owner, the income from the licensing of naming rights (and perhaps that from ancillary rights granted as part of the naming rights contract) represents the quantification of the economic benefit. The net present value of that economic benefit is the basis for quantifying value.

Estimating the value of a naming rights contract from the standpoint of a corporate lessee represents a much more complex situation. The value of naming rights is going to depend entirely on the nature of the rights being transferred. The value of the naming rights contract will depend on the extent to which it is *favorable* to the corporate lessee.

Cost Approach

In Chapter 7 we discussed the cost approach and how it is applied to intangible assets and intellectual property. Our conclusion was that it is rarely appropriate for intellectual property assets and that also holds true in the case of a naming rights contract. The cost of putting together such a contract or the cost of the facility involved is both irrelevant to the value of the deal. We say this even though it is not unlikely that the cost of renovating or constructing the facility may well have entered into the naming rights negotiations. That is, the facility owner, especially if it was a government-owned property, may well have "backed into" the payments that would be required to support an amount of debt that it was seeking for the renovation or construction cost. The amount of that annual payment, matched against the payment terms of the bonded debt, may well have heavily influenced the amount that the entity was seeking from the outside corporation for naming rights. This is coincidental to the question of value, however, and typically the value associated with a naming rights contract would have nothing to do with the cost of negotiating it or the cost of the facility or renovations that may have triggered the discussion.

Again, a real estate analogy is appropriate. When one constructs an office building that will be offered to the rental marketplace, the hoped-for rent should be enough to cover expenses, service the debt, and provide a return on the equity investment. In the end, however, the rental marketplace determines what the lease income will be. If construction costs were unusually high, the market-driven rent may not be enough to provide the hoped-for returns. We can bring this analogy back to the naming rights situation. The owner of a facility who faces substantial costs for construction or renovation may hope that the project can be financed entirely with debt whose service will be borne by the corporate partner.

It may be, however, that no corporation will come forward for such a transaction at the hoped-for expense. The owner must, therefore, scale back the expectations. There are, then, some limits on the amount that a corporate sponsor would be willing to pay, defined by the Principle of Substitution. That is, the corporation must weigh the cost of a naming rights contract against the cost of advertising and public relations benefits that could be obtained by other means. If the owner's construction costs were unusually low (or perhaps zero in the case of a pre-existing but unnamed facility), the owner is certainly not going to set rents at below-market levels even though they might be sufficient to provide a reasonable return on the investment.

Market Approach

As we touched on above, the marketplace transactions for naming rights have been very uneven, primarily because in the last few years there has been a rapidly surging interest in the leasing of naming rights by corporations and in the willingness of facility owners to lease naming rights. In addition, just as is the case with any intellectual property licensing activity, there is an almost infinite variety of terms and conditions associated with a naming rights transaction. Because of that, it is very difficult to observe value benchmarks in the marketplace.

From the basic data, a summary of which is in Appendix 17A, we examined some relationships that might provide some insight into valuation benchmarks from the marketplace. From the web site *foxsports.com* we extracted the team values (which we took to refer to the market value of the named teams) for the teams that occupy the facilities for which we had naming rights information. This data was attributed on the web site to *Forbes* magazine. Our supposition was that there would be some relationship between team value and the price of a naming rights contract at their "home" facility. This would seem logical in that team value (though we do not know the precise methodology used) would be related to the team success, the spectator size of its home field, and the size of the media market in which it performs at home. It seemed that these factors would also be elements of value for the naming rights in the facility. We first made this comparison using the team with the highest value (where there are two teams sharing a facility). This comparison is shown graphically in Figure 17A.1, below.

This analysis showed no observable relationship, so we made the comparison combining the value of both teams in the two-team facilities. That data is contained in Figure 17A.2.

Even if we compare total deal value with two-team values, no pattern emerges. If anything, the indication is that the more valuable the teams (in two-team facilities), the less the value of a naming rights deal (see Figure 17A.3).

It seems clear from this analysis that there are different elements that influence team value and naming rights value.

We then related naming rights contract term and value to the leagues of the incumbent team or teams. The longest contract terms were in evidence for Major League Baseball ("MLB"), followed by facilities in which National Basketball Association ("NBA") teams shared occupancy with National Hockey League ("NHL") teams. Contract term was about the same for minor league teams, NBA- and NHL-only facilities, and stadiums occupied by National Football League ("NFL") teams. These relationships are shown in Figure 17A.4.

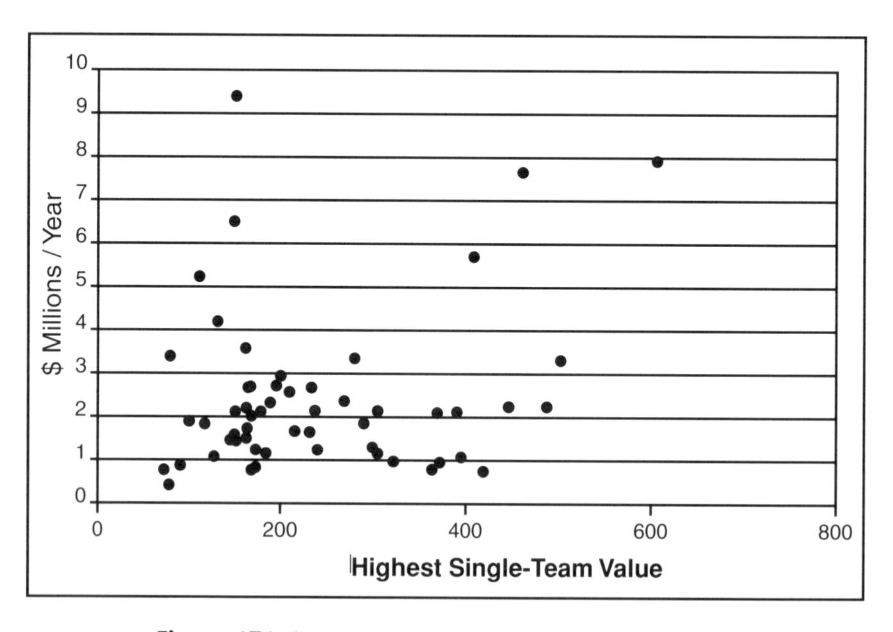

Figure 17A.1 Value per Year versus Single-Team Value

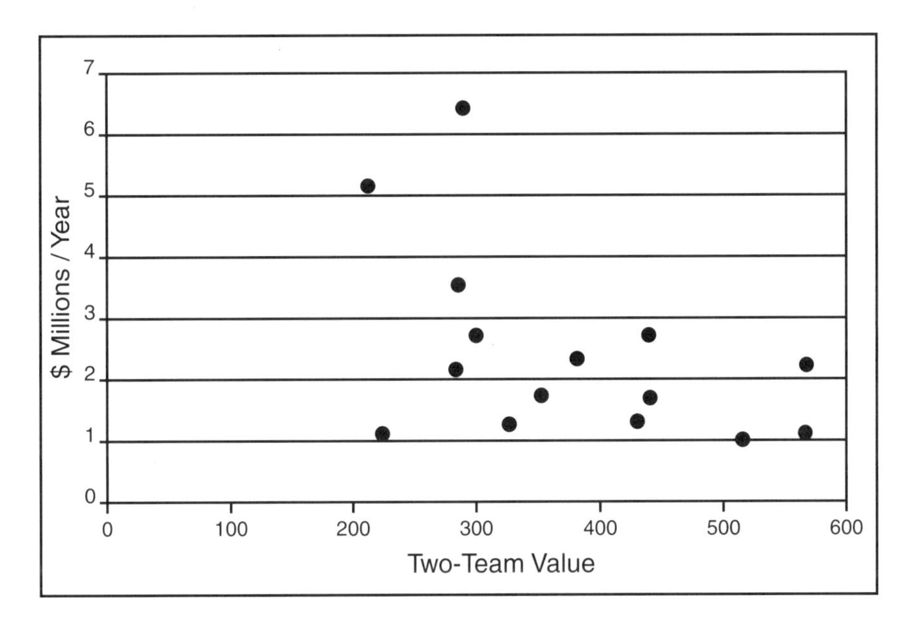

Figure 17A.2 Value per Year versus Two-Team Value

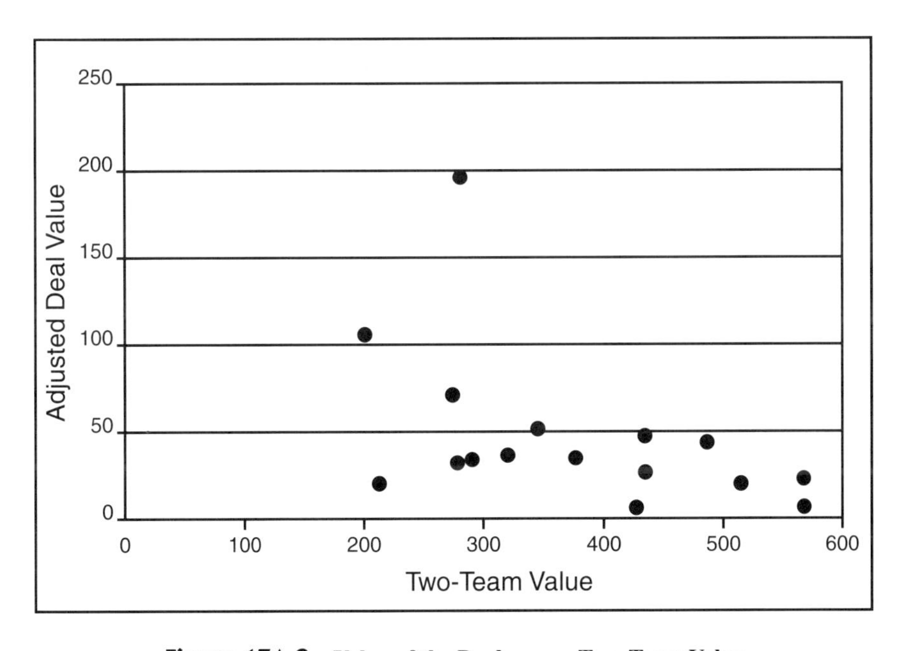

Figure 17A.3 Value of the Deal versus Two-Team Value

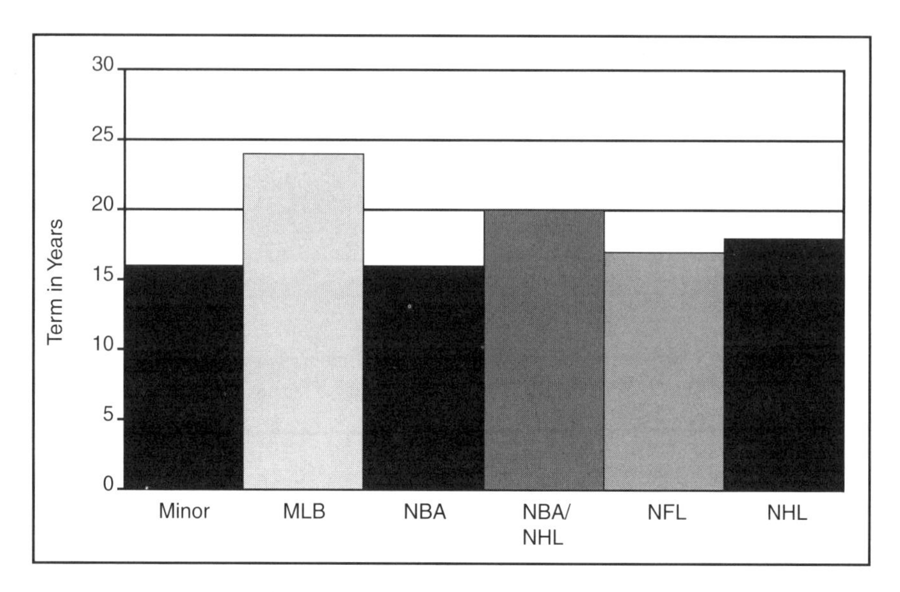

Figure 17A.4 **Deal Term by League**

There is more marked difference in the adjusted deal value by league, with NBA/NHL facilities commanding the highest deal values, followed by MLB and NFL facilities, with single-team facilities of the NHL and NBA following. Deal values for minor league facilities lag far behind (see Figure 17A.5).

While value relationships appear to be elusive, some elements are fairly clear. Annual deal value increases as the length of contract term increases (see Figure 17A.6).

Certainly there has been a sharp escalation of both annual payments and total deal value over the years. Figures 17A.7, 17A.8, and 17A.9 illustrate this.

We also observe that the range of deal values has widened. This would indicate that the contract terms vary considerably and that corporate lessees are looking carefully at the specifics of a naming rights contract and the potential economic benefits, rather than being driven by "the market" and seemingly escalating prices. We also wonder whether the prices of new deals are beginning to level off as the price of naming rights deals approaches the level of alternative advertising and public relations programs.

Whether or not we are correct, the message is clear that naming rights contracts must be valued based on their particular facts and circumstances. There is no overriding market data that can provide solid benchmarks.

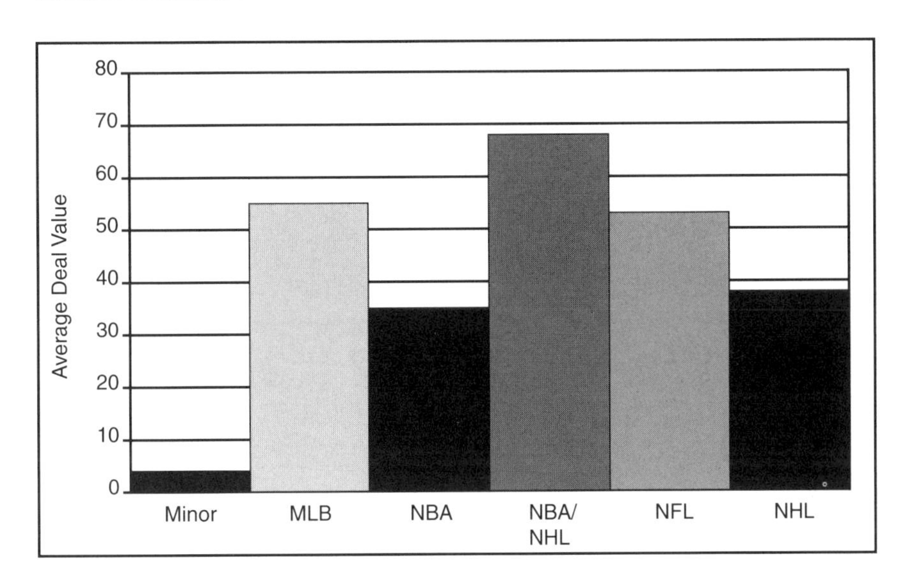

Figure 17A.5 Average Deal Value by League

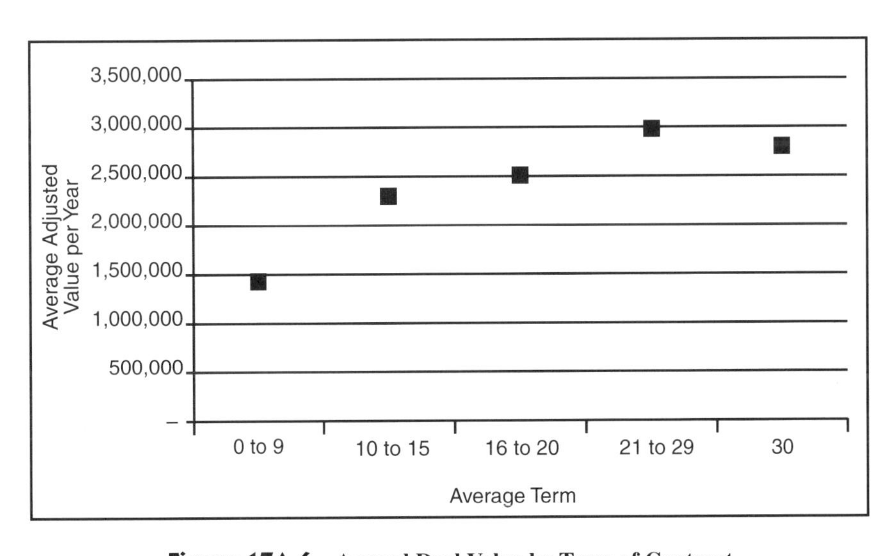

Figure 17A.6 Annual Deal Value by Term of Contract

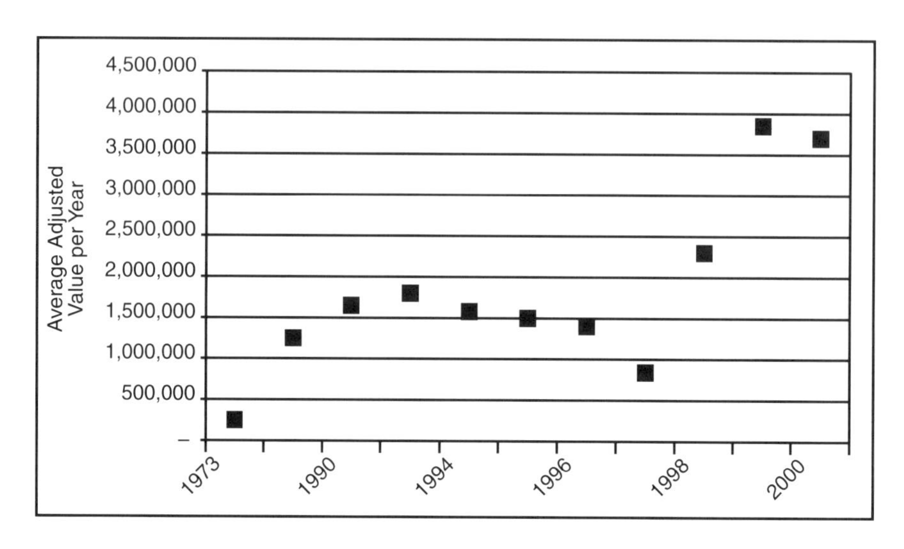

Figure 17A.7 Annual Deal Value by Contract Origination Date

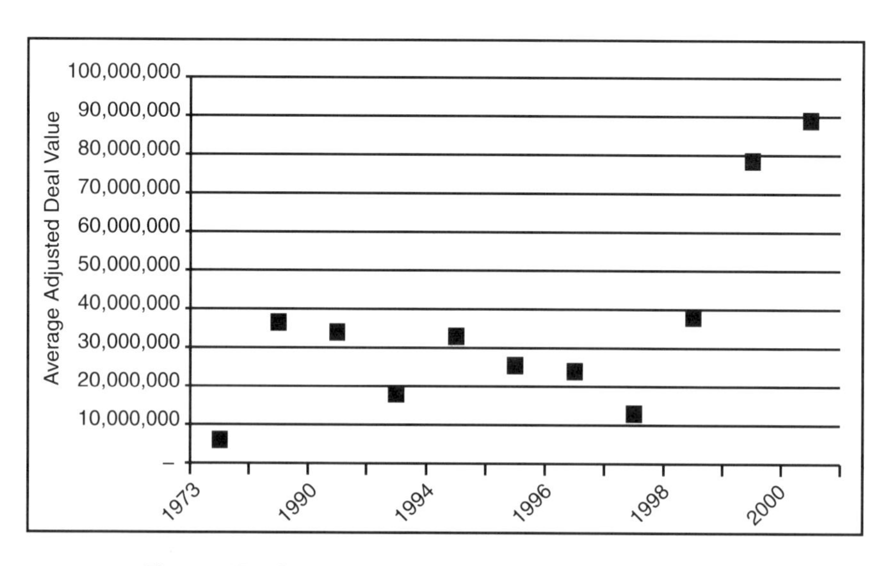

Figure 17A.8 Deal Value by Year of Contract Origination

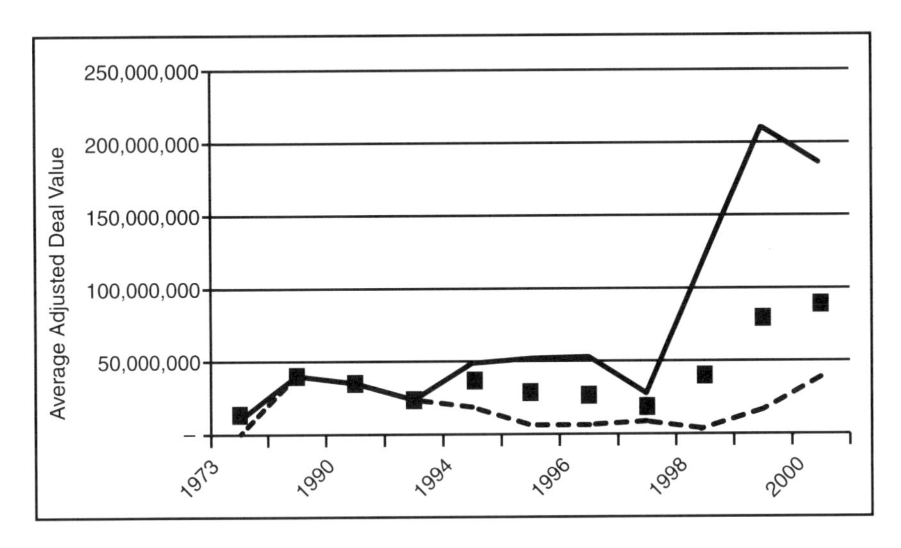

Figure 17A.9 **Deal Value by Year of Contract Origination, with High and Low Values Indicated**

Income Approach

The income approach for the valuation of intellectual property was discussed in detail in Chapters 8 through 10.

If our assignment is to value a naming rights contract from the standpoint of the facility owner (or other entity that has the right to collect all or a portion of the naming rights income stream), the task is fairly straightforward. There is a contractual agreement in place which calls for the payment of a certain amount of income annually from the appraisal date into the future. Some decision making must be made relative to the risk that the lessee of the naming rights will be able to honor that contract. Some judgment also must be made if there are options in the naming rights contract which permit the lessee to alter or cancel the contract based on certain events, such as the gain or loss of a major league team (or player) using the facility, physical damage to the facility, etc. This becomes a discounted cash flow exercise, which is described elsewhere in this book. The resulting value, of course, is the market value of the naming rights contract to the facility owner.

Much less straightforward is the valuation of a naming rights contract from the perspective of the corporate lessee. That is, the value will be measured by the extent to which the economic benefits (primarily of an enhanced advertising program) exceed their cost, as measured by the cost of equally effective advertising obtained in an alternative way. We believe that this is a controlling factor. A corporation "leases" naming rights primarily for the advertising benefit that they will provide. There are myriad advertising and public relations opportunities available. While each naming rights opportunity is unique, the overall objective (i.e., delivering advertising messages) is available by other means. Therefore, it seems reasonable to assume that the economic benefit that the lessee of naming rights would be willing to hand over to a facilities owner is likely going to be something less than the cost of equivalent advertising and public relations activities. To be sure, while competition for the naming rights to a facility or even corporate culture or management egos may drive the price near to or equal to the cost of alternative advertising, business logic would certainly suggest that the price would not go beyond that.

There can be other benefits in addition to those associated with advertising, of course, but in the overall we are seeking the difference between obtaining those benefits as part of the naming rights contract and obtaining them in some other way.

In applying the income approach, it is obvious that a simple discounted cash flow calculation would, on the surface, yield a negative result. The corporate lessee would be expected to experience an outflow of cash equal to the anticipated payments to the facility's owner under the contract. The offset to that, of course, is the present value of the costs that would be borne by the corporate lessee to obtain similar economic benefit by other means. These would be expenditures for advertising, public relations programs, promotion expenses, client entertainment expenses (at "retail"), test marketing programs, and the like. Each element of benefit provided by the naming rights contract should be analyzed as to the price at which it would be obtainable elsewhere. The difference between these two present value calculations indicates the value, to the corporate lessee, of the naming rights contract.

SUMMARY

Contracts for naming rights should be viewed in the context of "favorable contracts" as they have been discussed in previous chapters of this book. We must be careful to define the particular rights to be valued—be they those of the facility/owner lessor or the name owner/lessee. If we are attempting to predict the market value of a transaction, we have observed that there are few, if any, benchmarks from the marketplace. When this is the case, we must base the conclusion on our estimate of the economic benefits to be enjoyed by both parties, and how they would reasonably divide them.

Appendix 17A

Summary of Naming Rights Transactions

City	Facility Name	Naming Rights Lessee	Team 1	Team 2	League(s)	Year	Contract	Value		Value per Year(4)	
							Term (1)	Estimated(2)	Adjusted(3)	Estimated	Adjusted
San Francisco	3Com Park	3Com Corporation	San Francisco 49ers		NFL	1995	5	4.00	4.56	0.8	0.91
Nashville	Adelphia Coliseum	Adelphia Communications	Tennessee Titans		NFL	1999	15	30.00	31.20	2.00	2.08
Toronto	Air Canada Centre	Air Canada	Toronto Raptors	Toronto Maple Leafs	NBA/NHL	1999	15	30.40	31.62	2.03	2.11
Schaumburg	Alexian Field	Alexian Brothers Health Care	Schaumburg Flyers		Minor		10	2.00		0.20	
Rosemont	Allstate Arena	Allstate Insurance	Chicago Wolves	Chicago AFL	Minor		10	11.00		1.10	
Little Rock	ALLTEL Arena	Alltel Corporation	Arkansas RazorBlades		Minor		20	7.00		0.35	
Jacksonville	ALLTEL Stadium	Alltel Corporation	Jacksonville Jaguars		NFL	1995	10	6.20	7.07	0.62	0.71
Phoenix	America West Arena	America West Airlines	Phoenix Coyotes	Phoenix Suns	NBA/NHL	1989	30	26.00	36.40	0.87	1.21
Miami	American Airlines Arena	American Airlines	Miami Heat		NBA	1999	20	42.00	43.68	2.10	2.18
Dallas	American Airlines Center	American Airlines	Dallas Mavericks	Dallas Stars	NBA/NHL	2000	30	195.00	195.00	6.50	6.50
Sacramento	ARCO Arena	Atlantic Richfield Company	Sacramento Kings		NBA	1998	7	10.00	10.70	1.43	1.53
Sacramento	ARCO Arena	Atlantic Richfield Company	Sacramento Kings		NBA	1994	20	25.00	29.25	1.25	1.46
Anaheim	Arrowhead Pond of Anaheim	Arrowhead Mountain Spring Water	Mighty Ducks of Anaheim		NHL	1993	10	15.00	18.00	1.50	1.80
Sacramento	Atlantic Richfield Stadium	Atlantic Richfield Company	Sacramento Kings		Minor	1985	99	7.50	12.08	0.08	0.12
Memphis	AutoZone Park	AutoZone	Memphis Redbirds		Minor		25	4.30		0.17	
	Bank of America Center	Bank of America	Idaho Steelheads		Minor		10	1.00		0.10	
Phoenix	Bank One Ballpark	Bank One	Arizona Diamondbacks		MLB	1998	30	66.00	70.62	2.20	2.35
Chattanooga	BellSouth Park	Bell South	Chattanooga Lookouts		Minor		10	1.00		0.10	
Greenville	BI-LO Center	BI-LO	Greenville Grrrowl		Minor		10	3.00		0.30	
Rochester	Blue Cross Arena	Blue Cross	Rochester Americans		Minor		15	2.97		0.20	
Calgary	Canadian Airlines Saddledome	Canadian Airlines	Calgary Flames	Calgary Hitmen	NHL	1996	20	6.75	7.49	0.34	0.37
Cincinnati	Cinergy Field	Cinergy Corporation	Cincinnati Bengals	Cincinnati Reds	NFL/MLB	1998	6	6.00	6.42	1.00	1.07
Foxboro	CMGI Field	CMGI Corp.	New England Patriots		NFL	2000	15	114.00	114.00	7.60	7.60
Myrtle Beach	Coastal Federal Field	Coastal Federal S&L	Myrtle Beach Pelicans		Minor		10	1.00		0.10	
College Park	Comcast Center	Comcast Inc	Univ. MD		School	2000	25	20.00	20.00	0.80	0.80
Detroit	Comerica Park	Comerica Bank	Detroit Tigers		MLB	2000	30	86.00	86.00	2.87	2.87
Bridgewater	Commerce Bank Park	Commerce Bank	Unknown		Minor		15	3.55		0.24	
Houston	Compaq Center	Compaq Computer	Houston Rockets	Houston Aeros	NBA	1997	8	5.40	5.83	0.68	0.73

Indianapolis	Conseco Fieldhouse	Conseco Inc.	Indiana Pacers		NBA	1999	20	40.00	41.60	2.00	2.08
East Rutherford	Continental Airlines Arena	Continental Airlines	New Jersey Nets	New Jersey Devils	NBA/NHL	1996	12	29.00	32.19	2.42	2.68
Ottawa	Corel Centre	Corel Corporation	Ottawa Senators		NHL	2000	30	100.00	100.00	3.33	3.33
Round Rock	Dell Diamond	Dell Computer	Round Rock Express		Minor		15	2.50		0.17	
Salt Lake City	Delta Center	Delta Air Lines	Utah Jazz		NBA	1990	20	25.00	33.25	1.25	1.66
Atlanta	Discover Mills Mall	Discover Financial Services	Shopping Mall		Mall	2000	10	10.00	10.00	1.00	1.00
Buffalo	Dunn Tire Park	Dunn Tire Co.	Buffalo Bisons		Minor		6	1.58		0.26	
Long Island	EAB Park	EAB Bank	L.I. Ducks		Minor		10	2.30		0.23	
Anaheim	Edison International Field of Anaheim	Edison International	Anaheim Angels		MLB	1998	20	50.00	53.50	2.50	2.68
Houston	Enron Field	Enron Corporation	Houston Astros		MLB	2000	30	100.00	100.00	3.33	3.33
Charlotte	Ericsson Stadium	Ericsson Inc.	Carolina Panthers		NFL	1996	10	20.00	22.20	2.00	2.22
Washington	FedEx Field	Federal Express	Washington Redskins		NFL	1999	27	205.00	213.20	7.59	7.90
	Fieldcrest Cannon Stadium	Fieldcrest	Piedmont Boll Weevils		Minor		15	0.30		0.02	
Dayton	Fifth Third Field	Fifth Third Bank	Dayton Dragons		Minor		20	4.50		0.23	
Wilkes-Barre	First Union Arena	First Union Bank	W/S Penguins		Minor		10	2.30		0.23	
Philadelphia	First Union Center	First Union Corporation	Philadelphia 76ers	Philadelphia Flyers	NBA/NHL	1998	10	25.00	26.75	2.50	2.68
Philadelphia	First Union Center	First Union Bank	Philadelphia 76ers	Philadelphia Flyers	NBA/NHL	1994	29	40.00	46.80	1.38	1.61
Boston	Fleet Center	Fleet Financial Group	Boston Celtics	Boston Bruins	NBA/NHL	1995	15	30.00	34.20	2.00	2.28
St. Petersburg	Florida Power Park	Florida Power Corp.	St. Petersburg Devil Rays		Minor		10	3.00		0.30	
Salt Lake City	Franklin-Covey Field	Franklin-Covey	Salt Lake Buzz		Minor		10	1.40		0.14	
Rochester	Frontier Field	Frontier Telephone	Rochester Red Wings		Minor		20	3.50		0.18	
Nashville	Gaylord Entertainment Center	Gaylord Entertainment	Nashville Predators		NHL	1999	20	80.00	83.20	4.00	4.16
Vancouver	General Motors Place	General Motors	Vancouver Canucks	Vancouver Grizzlies	NBA/NHL	1995	20	18.50	21.09	0.93	1.05
Cleveland	Gund Arena	Gordon & George Gund	Cleveland Cavaliers	Cleveland Lumberjacks	NBA	1996	20	14.00	15.54	0.70	0.78
Chicago	Hawkinson Ford Field	Hawkinson Ford	Cook County Cheetahs		Minor		10	0.35		0.04	
Tampa Bay	Houlihan's Stadium	Houlihan's	Unknown		Minor		30	10.00		0.33	
Cleveland	Jacobs Field	Richard Jacobs	Cleveland Indians		MLB	1996	20	13.90	15.43	0.70	0.77
Ottawa	JetForm Park	JetForm	Ottawa Lynx		Minor		15	0.99		0.07	
Modesto	John Thurman Field	Modesto Bee	Modesto A's		Minor		10	0.25		0.03	
Seattle	Key Arena	Key Corporation	Seattle Supersonics	Seattle Thunderbirds	NBA	1995	15	15.10	17.21	1.01	1.15
Louisville	Louisville Slugger Park	Louisville Slugger Sports	Louisville RiverBats		Minor		10	2.00		0.20	
Charlotte	Lowe's Motor Speedway	Lowe's	N.A.		N.A.		10	35.00		3.50	
Buffalo	Marine Midland Arena	Marine Midland Bank	Buffalo Sabres	Buffalo Blizzard	NHL	1996	20	15.00	16.65	0.75	0.83

City	Facility Name	Naming Rights Lessee	Team 1	Team 2	League(s)	Year	Contract Term (1)	Value Estimated(2)	Value Adjusted(3)	Value per Year(4) Estimated	Value per Year(4) Adjusted
Washington	MCI Center	MCI Communications Corporation	Washington Wizards	Washington Capitals	NBA/NHL	1995	20	44.00	50.16	2.20	2.51
Pittsburgh	Mellon Arena	Mellon Financial Corporation	Pittsburgh Penguins		NHL	1999	10	18.00	18.72	1.80	1.87
Lynchburg	Merritt Hutchinson Stadium	Merritt Hutchinson	Lynchburg Hillcats		Minor	2000	10	1.00	1.00	0.10	0.10
Milwaukee	Miller Park	Miller Brewing	Milwaukee Brewers		MLB	2000	20	41.20	41.20	2.06	2.06
Winnipeg	Mind Field	Mind Computer	Unknown		Minor		10	1.00		0.10	
Montreal	Molson Centre	Molson Companies	Montreal Canadiens		NHL	1998	30	33.10	35.42	1.10	1.18
Miami	National Car Rental Center	National Car Rental	Florida Panthers		NHL	1998	10	25.00	26.75	2.50	2.68
Oakland	Network Associates Coliseum	Network Associates	Oakland Raiders	Oakland Athletics	NFL/MLB	1998	5	6.00	6.42	1.20	1.28
Fargo	Newman Outdoor Field	Newman Outdoor	Fargo-Moore. Redhawks		Minor		5	5.00		1.00	
Buffalo	North Americare Park	Americare	Buffalo Bisons		Minor		13	3.90		0.30	
Lansing	Oldsmobile Park	Oldsmobile	Lansing Lugnuts		Minor		10	1.50		0.15	
Syracuse	P&C Stadium	P&C Grocers	Syracuse SkyChiefs		Minor		25	3.50		0.14	
San Francisco	Pacific Bell Park	Pacific Bell	San Francisco Giants		MLB	2000	24	50.00	50.00	2.08	2.08
Victoria	Pacific Coast Net Place	Pacific Coast Net	Victoria Spiders		Minor		15	5.96		0.40	
Louisville	Papa John's Cardinal Stadium	Papa John's Pizza	Unknown		Minor		10	5.00		0.50	
Albany	Pepsi Arena	PepsiCo	Albany Firebirds	Albany River Rats	Minor		10	8.00		0.80	
Denver	Pepsi Center	PepsiCo	Denver Nuggets	Colorado Avalanche	NBA/NHL	1999	20	68.00	70.72	3.40	3.54
Indianapolis	Pepsi Coliseum	PepsiCo	Indianapolis Ice		Minor		5	0.80		0.16	
Portland	PGE Park	Portland General Electric	Unknown		Minor		10	7.10		0.71	
Atlanta	Philips Arena	Royal Philips Electronics	Atlanta Hawks	Atlanta Thrashers	NBA/NHL	1999	20	180.00	187.20	9.00	9.36
Pittsburgh	PNC Park	PNC Bank	Pittsburgh Pirates		MLB	2001	20	30.00	30.00	1.50	1.50
	Pringles Park	P&G	West Tennessee Dia.Jaxx		Minor		15	1.00		0.07	
Miami	Pro Player Stadium	Fruit of the Loom	Miami Dolphins	Florida Marlins	NFL/MLB	1996	10	20.00	22.20	2.00	2.22
Baltimore	PSINet Stadium	PSINet	Baltimore Ravens		NFL	1998	20	105.50	112.89	5.28	5.64
San Diego	Qualcomm Stadium	Qualcomm Inc.	San Diego Chargers	San Diego Padres	NFL	1997	20	18.00	19.44	0.90	0.97
Tampa Bay	Raymond James Stadium	Raymond James Financial	Tampa Bay Buccaneers	Tampa Bay Mutiny	NFL	1998	18	55.00	58.85	3.06	3.27
Indianapolis	RCA Dome	RCA	Indianapolis Colts		NFL	1994	10	10.00	11.70	1.00	1.17
Erie	Rich Stadium	Rich Products Corporation	Buffalo Bills		NFL	1973	25	1.50	5.87	0.06	0.23

Staten Island	Richmond County Bank Ballpark	Richmond County Bank	Staten Island Yankees		Minor	2000	9	3.60	3.60	0.40	0.40
Jupiter	Roger Dean Stadium	Roger Dean Chevrolet	Jupiter Hammerheads		Minor		10	1.00		0.10	
Seattle	Safeco Field	Safeco Corporation	Seattle Mariners		MLB	1999	20	36.00	37.44	1.80	1.87
St. Louis	Savvis Center	Savvis Communications	Unknown		NHL	2000	20	70.00	70.00	3.50	3.50
San Antonio	SBC Center	SBC Communications	San Antonio Spurs		NBA	2000	20	41.00	41.00	2.05	2.05
Green Bay	ShopKo Hall	ShopKo	Unknown		Minor		20	1.40		0.07	
Edmonton	Skyreach Centre	Skyreach Equipment	Edmonton Oilers	Edmonton Drillers	NHL	1998	5	3.40	3.64	0.68	0.73
Trenton	Sovreign Bank Arena	Sovreign Bank	Trenton Titans	Trenton Stars	Minor		10	2.68		0.27	
Los Angeles	STAPLES Center	Staples	Los Angeles Clippers	Los Angeles Kings	NBA/NHL	1999	20	100.00	104.00	5.00	5.20
Minneapolis	Target Center	Target	Minnesota Timberwolves		NBA	1995	15	18.75	21.38	1.25	1.43
Orlando	TD Waterhouse Center	TD Waterhouse	Orlando Magic		NBA		5	7.80		1.56	
Estero	TECO Arena	TECO Energy	Florida EverBlades		Minor		20	7.00		0.35	
Edmonton	TELUS Field	TELUS Corp.	Edmonton Trappers		Minor		20	1.32		0.07	
Sacramento	Thomas P. Raley Field	N.A.	Sacramento River Bats		Minor		20	15.00		0.75	
St. Louis	Trans World Dome	Trans World Airlines	St. Louis Rams		NFL	1995	20	36.70	41.84	1.84	2.09
Tampa Bay	Tropicana Field	Tropicana Dole Beverages	Tampa Bay Devil Rays		MLB	1996	30	46.00	51.06	1.53	1.70
Tucson	Tuscon Electric Park	Tuscon Electric Co.	Tuscon Sidewinders		Minor		10	2.00		0.20	
Chicago	United Center	United Airlines	Chicago Bulls	Chicago Blackhawks	NBA/NHL	1994	20	36.00	42.12	1.80	2.11
Grand Forks	Unknown	First National Bank	Unknown		Minor		20	3.00		0.15	
Rochester	Unknown	Paetec Corp.	Unknown		Minor	1973	25	1.50	5.87	0.06	0.23
Jefferson County	US West Stadium	US West Communications	Unknown high school		School	2000	10	2.00	2.00	0.20	0.20
Tucson	Wells Fargo Arena	Wells Fargo Bank	Unknown		Minor		10	5.00		0.50	
St. Paul	Xcel Energy Center	Excel Energy Corp.	Unknown		Minor		25	75.00		3.00	

NOTES:

Source: This information is from a variety of sources, including the web site *www.sportsvenues.com* of Mediaventures, Milwaukee, WI, publisher of "Revenues from Sports Venues," the *www.foxsports.com* web site of News Digital Media, Inc., New York, NY, "Naming Rights Deals" published by Team Marketing Report, Inc., Chicago, IL, various news releases, newspaper articles.

(1) Term is in years.

(2) Estimated value is the total value of the transaction.

(3) Estimated value adjusted to current level by means of the CPI.

(4) Estimated value divided by the term in years.

Appendix F

Valuation Resources (New)

This new appendix presents resources that can supply some of the puzzle pieces that are needed to complete a valuation of intellectual property and intangible assets. We have used many of these resources and can recommend them to you. Included in this appendix are publications, Internet services, professional associations, and databases. The information provided here is not an exhaustive list of all possible resources; it is simply a great place to start investigations.

The following list focuses on some of the key attributes of conducting a valuation. For example, when using an income approach for valuing intellectual property, the relief-from-royalty approach is often considered, as described in Chapter 8 of the main text on page 222. A critical component for implementing this method is a royalty rate that represents the savings enjoyed by a company from owning a particular intellectual property. Market transaction data is very scarce regarding royalty rates. This appendix identifies sources for finding such information.

Other information provided here includes publications that describe the valuation process and sources of information for development of discount rates. The discount rate reflects the risk associated with investment in a particular intellectual property and is required for implementation of the income approach. Resources are identified that can help you identify the components needed to develop a discount rate.

General background information about companies, their competitors, and their industries are also needed for valuations. The first place to start is the web site of the Securities and Exchange Commission, where detailed financial information about every publicly traded company is available.

Also discussed in this appendix is the Licensing Executives Society. This professional organization is the preeminent organization for learning about intellectual property and how it is implemented into business strategies. Membership is highly recommended for anyone wishing to establish a career in any aspect of intellectual property.

ROYALTY RATE INFORMATION

Presented below are sources of information for royalty rates from real-world market transactions. The sources presented include:

- Intellectual Property Research Associates
- RoyaltySource®
- The Financial Valuation Group
- Licensing Economics Review

Intellectual Property Research Associates

IPRA was founded in 1993 by Russell L. Parr, CFA, ASA, to research the value of intellectual property including patents, trademarks, and copyrights. The company has gathered an impressive amount of information on royalty rates and intellectual property values. The information in their publications includes details about the companies involved in a particular transaction as well as a description of the intellectual property that was transferred and the financial terms associated with each reported transaction. The results of their research are available to you in the various reports listed below. The reports offered by IPRA are:

- *Royalty Rates for Trademarks and Copyrights,* 2nd Edition
- *Royalty Rates for Technology,* 2nd Edition
- *Royalty Rates for Pharmaceuticals and Biotechnology,* 4th Edition

All three of these books can be ordered directly from *ipresearch.com.*

***Royalty Rates for Trademarks and Copyrights,* 2nd Edition**

Too often people think of T-shirts, caps, or key chains when they hear about licensing transactions. Too often they think only of trinkets and trash. However, this old-fashioned approach to licensing is increasingly outdated. Licensing has become the ultimate marketing strategy, and the approach to licensing and merchandising has changed dramatically in the last 10 years. Increasingly, corporate America and European companies think of licensing and merchandising as part of a longer-term strategic commitment, rather than a short-term approach for increasing revenues. In addition to royalty income, licensing provides significantly increased consumer awareness that could only be otherwise obtained from increases to already hugely expensive advertising campaigns. Trademark and copyright royalty rate information is provided in this book for companies in the following industries:

Airline
Apparel
Architecture
Art
Boats
Celebrities
Communications
Corporate Names
Electronics
Food
Franchises
Furniture
General Merchandise
Goods
Medical
Movies
Music
Party
Publishing
Restaurants
Sports
Toys
University Names

***Royalty Rates for Technology,* 2nd Edition**

This new report gives you information about technology royalty rates. It also shows other measures of technology value including license fees and milestone payments. *Royalty Rates for Technology* gives you the information you need to negotiate valuable license agreements. The trend for royalty rates continues upward. Large corporations are looking at their intellectual property portfolios as key assets that deserve specialized management. They are establishing subsidiaries with the sole purpose of managing and licensing their technology. Many other companies are completely dependent on their technology for continued survival in the marketplace, and these forces are driving the royalty rates to new levels. *Royalty Rates for Technology* will let you see the new levels to which royalty rates are rising. Royalty rate information is provided for technology transfers that have happened in the following industries:

Aeronautics	Computer Hardware	Entertainment	Household Products	Semiconductors
Agriculture	Computer Software	Financial	Mechanical	Sports
Automotive	Construction	Food	Medical	Steel
Chemistry	Electrical	Franchises	Natural Resources	Toys
Communications	Electronics	Glass	Photography	Waste Treatment

Royalty Rates for Pharmaceuticals and Biotechnology, 4th Edition

The *Royalty Rates for Pharmaceuticals and Biotechnology* is a comprehensive tool to help you maximize the value of biotechnology and pharmaceutical technology. It shows how to price technology for licensing and strategic alliances.

Part One presents the theory of quantifying technology value and royalty rates for use in transferring technology. This section introduces a business framework to use as the foundation of technology valuation, then reviews the most commonly used royalty rate derivation methods, and explains the strengths and weaknesses of each. Subsequent chapters demonstrate how to estimate the investment risk associated with different stages of technology and use investment rate of return analysis to value technology and derive royalty rates.

The second part of this report presents detailed financial information about third-party transactions that center on the transfer of biotechnology and pharmaceutical technology. The players are identified, the technology is described, and all of the financial terms available are reported. Details are reported for License Agreements and Strategic Alliances.

RoyaltySource

Whether for negotiation, valuation, or infringement damage measurement, this division of AUS Consultants has been investigating and tracking royalty rate information from arm's-length licensing transactions for over 15 years. The result of this continuous investigation has yielded a searchable database of documented technology and trademark sale and licensing transactions that can minimize the time spent to research the marketplace for this information.

RoyaltySource continues to research all forms of media for reported transactions. Its intellectual property transaction database includes:

- Licensee and Licensor, including industry description or code
- Description of the property licensed or sold
- Royalty rate details
- Other compensation, such as upfront payments or equity positions
- Transaction terms, such as exclusivity, geographical restrictions, or grant-backs
- Source of Information

Customized searches are provided from a consultant who works with you to find exactly what you need. Access to this database is via ***royaltysource.com.***

The Financial Valuation Group

The Financial Valuation Group is an Internet site that provides a broad range of services to those involved in the appraisal professional. The services are primarily directed at professionals who are focusing on the valuation of businesses. However, they have developed a proprietary database of empirical research on intellectual property. This research is a compilation of intellectual property transactions gleaned from publicly available documents. Industries covered include sporting goods, software, pharmaceuticals, apparel, medical, restaurants, and telecommunications. The database is searchable by SIC code or NAICS code. Reports on individual transactions can be purchased online, but the identities of the licensor and licensee are not provided. Information can be obtained at *fvgi.com.*

Licensing Economics Review

This bi-monthly newsletter, published by AUS Consultants and edited by Russell L. Parr, reports intellectual property transaction data for all forms of intellectual property. Transactions involving licenses, gifts, outright sales, and strategic alliances are reported. The information in each issue includes details about the companies involved in a particular transaction as well as a description of the intellectual property that was transferred and the financial terms associated with each reported transaction. The newsletter also reports the financial terms of infringement litigation damages, awards, and settlements. Contact Beth McAndrews at (856) 243-1199, AUS Consultants, for a free examination issue of this publication.

DEVELOPMENT OF A DISCOUNT RATE

The best source for information about discount rates and the history of investment rates of return is Ibbotson Associates. An overview of the information that is available from this company is presented in this section of the appendix.

Ibbotson Associates

Ibbotson Associates, established by Professor Roger Ibbotson in 1977, offers consulting and training services as well as software, data, and presentation products regarding investment rates of return that can be used to develop discount rates for income approach valuations.

They provide solutions to investment and finance problems for a diverse set of markets, ranging from financial planners and brokers to large investment managers and corporations. Ibbotson Associates fills a growing need in the finance industry as a single-source provider of investment knowledge, expertise, and technology.

Ibbotson Stocks, Bonds, Bills, and Inflation Yearbook is the definitive study of historical U.S. capital markets data. Each publication is composed of invaluable information and represents the industry standard. Their comprehensive collection covers such topics as domestic and international historical data, Treasury yield curve statistics, and cost of capital by industry.

Ibbotson Associates has been providing the investment industry with revolutionary software and data for more than a decade. Today, the firm continues to utilize cutting-edge technology to provide end-users with easy-to-use, graphical software. Their unique approach has earned Ibbotson Associates the reputation as a leading provider of innovative investment tools in the financial industry. Some of their products include:

> *Cost of Capital Quarterly* includes industry cost of capital analysis on over 300 industries for help in performing discounted cash flow analysis. Industry betas, cost of equity, weighted average cost of

capital, and other important financial statistics are presented by industry. Industry analysis by SIC code can be purchased online.

Stocks, Bonds, Bills, and Inflation: Valuation Edition is an easy-to-understand overview and comparison of the build-up method, CAPM (Capital Asset Pricing Model), Fama-French 3-factor model, and the DCF (discounted cash flow) approach. Tables are provided that enable you to calculate equity risk premia and size premia for any time period. Also provided are: evidence of size premia by industry; alternative methods of calculating equity risk premia, size premia, and beta; new developments in the field of cost of capital estimation; and problems and possible solutions in estimating the cost of capital for international markets.

Cost of Capital: Research Articles covering subjects such as equity risk premiums, size premium, betas, and minority discount rates. Available free online.

Company Analysis: Betas contains traditional 60-month beta calculations, betas for thinly traded securities, levered and unlevered betas, and betas adjusted toward peer group averages instead of toward the market.

International Cost of Capital Report provides cost of equity estimates for more than 130 countries.

GENERAL INFORMATION

When we need information about the health of the U.S. economy, we go to STAT-USA. When we need detailed information about publicly traded companies, we visit the EDGAR site of the Securities and Exchange Commission and Market Guide Inc.

STAT-USA

STAT-USA, a service of the *U.S. Department of Commerce,* is the site for the U.S. business, economic, and trade community, providing authoritative information from the federal government. Access to this web site is via *Stat-USA.gov.* Two of its premiere services include:

State of the Nation—Access this area for current and historical economic and financial releases and economic data. Stay informed with direct access to the federal government's wealth of information on the U.S. economy. Access to these files is provided for subscribers only.

GLOBUS & NTDB—Access this area for current and historical trade-related releases, international market research, trade opportunities, country analysis, and the trade library.

Also posted on this site are the speeches and comments of Federal Reserve Chairman Alan Greenspan. Regularly, Mr. Greenspan testifies before Congress about the health and prospects of the nation. His speeches are a succinct description of the U.S. economy.

SEC—EDGAR

The place to go for companies' 10-Ks and other SEC filings is the EDGAR site of the Securities and Exchange Commission. At this location, you will find access to the filings of all public companies. EDGAR (Electronic Data Gathering, Analysis, and Retrieval system) performs automated collection, validation, indexing, acceptance, and forwarding of submissions by companies and others who are required by law to file forms with the U.S. Securities and Exchange Commission (SEC). Its primary purpose is to increase the efficiency and fairness of the securities market for the benefit of investors, corporations, and the economy by accelerating the receipt, acceptance, dissemination, and analysis of time-sensitive corporate information filed with the agency.

Not all documents filed with the Commission by public companies will be available on EDGAR. Companies were phased into EDGAR filing over a three-year period, ending May 6, 1996. As of that date, all public domestic companies were required to make their filings on EDGAR, except for filings made in paper because of a hardship exemption. Third-party filings with respect to these companies, such as tender offers and Schedules 13D, are also filed on EDGAR.

Some documents are not yet permitted to be filed electronically and, consequently, will not be available on EDGAR. Other documents may be filed on EDGAR voluntarily and, consequently, may or may not be available on EDGAR. For example, Forms 3, 4, and 5 (security ownership and transaction reports filed by corporate insiders), and Form 144 (notice of proposed sale of securities) may be filed on EDGAR at the option of the filer. Similarly, filings by foreign companies are not required to be filed on EDGAR, but some of these companies do so voluntarily. (Note: Until recently, this was also the case with Form 13F, the reports filed by institutional investment managers showing equity holdings by accounts under their management. However, on January 12, 1999, the SEC released a rule to require electronic filing of the form as of April 1, 1999.)

It should also be noted that the actual annual report to shareholders (except in the case of investment companies) need not be submitted on EDGAR, although some companies do so voluntarily. However, the annual report on Form 10-K or 10-KSB, which contains much of the same information, is required to be filed on EDGAR.

Filers may choose to accompany their official filings with a copy in PDF. In order to read a document saved as a PDF, you need Adobe Acrobat reader. Direct access to this site can be found at *sec.gov/edaux/searches.htm.*

Publications

Listed below are some of the books that we have in our library along with the publishers' description of each book.

Trademark Valuation, Gordon V. Smith, John Wiley & Sons, 1996

Trademarks are among the most intangible of assets, yet they can have enormous value for an enterprise. The pink color of Owens-Corning insulation, McDonald's golden arches, the unique shape of the classic Coke bottle, these words, symbols, and colors embody the goodwill of the companies and institutions they represent. Potent cultural icons, trademarks are associated with quality, security, and even a sense of belonging in the minds of consumers. But how, exactly, do you determine the value of your trademark? How do you know if you are getting the best return on investment from your trademark? And what are the potential advantages and disadvantages of licensing your trademark, or even selling it outright?

The first guide devoted exclusively to an increasingly important area of intellectual property, *Trademark Valuation* provides answers to these and all your questions about how to value your trademark and to develop strategies for exploiting its full potential.

Gordon V. Smith, a consultant with more than three decades of experience advising clients on the value of their intellectual property, dispels common myths and misconceptions about trademarks and replaces them with logical, down-to-earth, practical guidance. Employing his unique talent for translating complex legal and financial concepts into plain English, he acquaints you with all the key legal and financial concepts, terms, principles, and practices, and guides you step-by-step through the entire valuation process. And, perhaps most importantly, he shows you how to use the information derived from your valuation to develop surefire strategies for getting the most out of your trademark.

With the help of dozens of case studies, Smith places the subject of trademark management in a contemporary, real-world context. He examines the role of crucial factors such as trademark longevity and offers guidelines for analyzing current and future market trends. He explores the implications of the emerging world marketplace. And he considers various worst-case scenarios, including infringement and piracy, bankruptcy, acts of consumer terrorism, and other potential crises that can have a disastrous effect on the value of a trademark.

Trademark Valuation is required reading for valuation experts, trademark specialists, and licensing executives, as well as the accountants and attorneys who work with them. It is also a valuable reference for advertising executives, business appraisers, and institutional investors.

Intellectual Property: Licensing and Joint Venture Profit Strategies, 2nd Edition, Gordon V. Smith and Russell L. Parr, John Wiley & Sons, 1998

Companies are increasingly looking to their intellectual property (patents, trademarks, formulas, copyrights, brand names, distributions systems, etc.) as a profit center. As they try to extract more value from their holdings, some of which have been left dormant for years, many are looking beyond their own core products to partnerships with outside industries. Intellectual property owners need to know how to exploit their product to the fullest extent.

Early-Stage Technologies: Valuation and Pricing, Richard Razgaitis, John Wiley & Sons, 1999

This popular book is a complete guide to technology risk management, valuation, and pricing. It shows how to identify key early-stage technologies and determine the value to individual companies, as well as provides methods for pricing pre-commercial products for sale or licensing. Topics include methods of valuation, the identification of risk factors, sources of value, the psychology of buying and selling, equity realizations, and negotiation strategies. Written by a professional who has spent his career making business decisions about embryonic technology investments, this book is definitely worth reading.

The Valuation of Technology: Business and Financial Issues in R&D, F. Peter Boer, John Wiley & Sons, 1999

"*The Valuation of Technology* is a timely and thoughtful book on a critical issue in the global business arena. Peter Boer's insights constitute important reading for leaders in all fields."—Jeffrey E. Garten, Dean, Yale School of Management

"*The Valuation of Technology* fills a critical void for those executives who wish to upgrade technology decision making from an art to a more definable science."—George B. Rathmann, Chairman and CEO, ICOS Corporation

Technology valuation has replaced risk management as the management approach to analyzing the profitability of current and future technology projects.

The Valuation of Technology: Business and Financial Issues in R&D explores the link between research and development and shareholder value in a comprehensive way, providing mathematical models for the valuation of R&D projects and answering critical questions on how to analyze technology initiatives and forecast their future value. This professional reference creates a common language for understanding the financial issues relating to R&D and provides analytical tools that businesspeople, scientists, and engineers can use to assess new technologies, R&D projects, and R&D budgets—thereby facilitating communication and producing more enlightened decisions. It also identifies several common fallacies in performing valuation of technology-based properties, including adding together enterprises with different time horizons and failing to recognize the value of risk-minimization strategies. Among the many remarkable features of *The Valuation of Technology* are that it offers quick, easy models for technology valuation that readers can use immediately; includes a method for the quantitative valuation of technology projects; shows readers how to build a project spreadsheet and assign value to research projects; and comes with a disk containing templates for a selection of mathematical models provided in the book.

***Value Driven Intellectual Capital: How to Convert Intangible Corporate Assets into Market Value,* Patrick H. Sullivan, John Wiley & Sons, 2000**

Intellectual capital provides a significant competitive advantage for companies. Intangible assets—product innovation, patents, copyrights, know-how, and corporate knowledge—have become as important as brick, mortar, and equipment. This informative reference provides strategies to meet the needs of those interested in the financial implications of intellectual capital. This book provides a corporate and financial executive's handbook to the new world of intangible assets and explains the new, boundary-expanding world of intellectual assets in which translating an innovative idea into bottom-line profits involves a tightly focused strategy with clear directives for making it happen.

***Technology Licensing: Corporate Strategies for Maximizing Value,* Russell L. Parr and Patrick H. Sullivan, John Wiley & Sons, 1996**

Russell Parr and Patrick Sullivan, along with a team of distinguished experts working at the front lines of technology licensing, reveal how today's top technology-based companies are maximizing the value of and return on their intellectual property. They also offer hands-on advice and guidance on how you can do the same in your company. With the help of numerous real-life case studies that demonstrate licensing strategies now used by DuPont, Xerox, Kodak, AlliedSignal, Hewlett-Packard, Dow Chemical, and other industry leaders, they tell you everything you need to know to:

- Determine where technology licensing best fits in your company's overall business strategies
- Establish a successful licensing program tailored to your company's vision and goals
- Create and successfully manage a technology portfolio
- Quickly and easily calculate royalty rates
- Put the lessons learned at top technology-based companies to work in your company

"Technology licensing strategies are now key instruments for accomplishing the corporate visions set forth by future-thinking companies. Look at any corporate mission statement and you will find the seeds of a strategy-based technology licensing program."—Russell Parr and Patrick Sullivan

In today's volatile, hypercompetitive global marketplace, cooperation and the sharing of intellectual property are keys to success. Of course, one of the most valuable forms of intellectual property is technology. More often than not, innovation and increased market penetration are the direct result of combining technologies from a variety of sources. Consequently, many companies have begun to devote more and more of their strategic efforts to discovering the best ways to manage technology so as to maximize value and return. For instance, AT&T has set up an independent business group to manage its intellectual property as a separate profit center, while other companies continue to run licensing through their legal and R&D departments. Which approach makes the most sense for your company, and why? Get the answers to these questions and many others in *Technology Licensing.*

Valuation: Measuring and Managing the Value of Companies, 3rd Edition, Tim Koller, Jack Murrin, and Tom Copeland, John Wiley & Sons, 2000

Hailed by financial professionals worldwide as the single best guide of its kind, *Valuation* provides crucial insights into how to measure, manage, and maximize a company's value. This long-awaited Third Edition has been updated and expanded to reflect business conditions in today's volatile global economy. In addition to all new case studies, it now includes in-depth coverage of real options and insurance companies, along with detailed instructions on how to drive value creation, and expert advice on how to manage difficult situations. It describes techniques for multibusiness valuations, valuation within an international context, and valuation for restructurings and mergers and acquisitions. It includes a companion web site featuring an interactive valuation-modeling application.

Written for those wanting to improve their ability to create value for the stakeholders in their businesses. It addresses estimating the value of alternative corporate and business strategies, assessing major transactions such as mergers, divestitures, recapitalizations, and share repurchases.

Internet Sites

Many web sites exist that provide good information about intellectual property. We like these two.

Franklin Pierce Law Center

Franklin Pierce Law Center, continuously ranked among the top five IP law schools in the United States, maintains a very informative web site called the Intellectual Property Mall *(ipmall.fplc.edu).* Through it one can access FPLC's IP library, publications and papers, and legal references such as the patent exams of the U.S. Patent and Trademark Office. This site also provides links to other worthwhile IP web sites. In its "Tools & Strategies" section, the Mall provides access to over 100 Briefing Papers covering a very wide range of patent, copyright, and trademark issues; case law; and practical assistance for IP practitioners.

This is a most useful and continually growing site.

Delphion Intellectual Property Network

The Delphion Intellectual Property Network (IPN) has evolved into a premier web site for searching, viewing, and analyzing patent documents. IPN provides you with free access to a wide variety of *data collections* and patent information including:

- United States patents
- European patents and patent applications
- PCT application data from the World Intellectual Property Office
- Patent Abstracts of Japan
- *INPADOC* family and legal status data
- *IBM Technical Disclosure Bulletins*

Searching is fast and easy. Along with simple keyword search, IPN offers alternative searches by patent number, boolean text, and advanced text that allows for multiple field searching. *Browsing* provides an organized approach to searching for patents. Through a review of specific classifications, you can identify topics and patents of interest. All collections are cross-referenced and forward and backward linked to all other referencing documents for immediate access to related information. Other standard features include:

- *Full text searching of U.S. patents*
- *Cross collection searching*
- *Access to business research about patent applicants*
- Hit-list sorting and management
- Query refinement
- Enhanced image viewing
- Access to the *IP Pages,* a directory of patent and IP-related resources on IPN

PROFESSIONAL ORGANIZATIONS

Here is information about the professional organization of which we are proudly members. This is the group to be with to learn about integrated intellectual property and business strategies.

Licensing Executives Society

The Licensing Executives Society (U.S.A. and Canada), Inc. is a professional society composed of more than 4,700 members engaged in the transfer, use, development, manufacture, and marketing of intellectual property. The membership covers a wide range of professionals, including business executives, lawyers, licensing consultants, engineers, academicians, scientists, and representatives of government. Membership is on an individual basis and is not transferable. Many major companies, professional firms, and universities are represented within the society's membership. LES (U.S.A. and Canada), Inc. is a member society of the Licensing Executives Society International, which has a worldwide membership of over 9,000 members in more than 25 national societies, representing over 60 countries. The main objectives of the organization are:

- To encourage high standards and ethics among persons engaged in domestic and international licensing.

- To hold meetings, seminars, and training courses for education and the exchange and dissemination of knowledge and information on licensing and intellectual property.
- To assist members in improving their skills and techniques.
- To inform the business community, public and governmental bodies of the economic significance and importance of licensing.
- To monitor domestic and international changes in the law and the practice of licensing and protecting intellectual property.
- To encourage the publication of articles, reports, statistics, and other materials on licensing and protecting intellectual property.

Some of the greatest benefits from this organization can be obtained by attending some of its annual conferences. Education and networking are the focal point of each conference.

Appendix G

Intellectual Property Management Institute (New)

If you are looking for a formal educational program in intellectual property management, this new institute is an excellent place to start.

The Intellectual Property Management Institute (IPM*i*) is a not-for-profit organization dedicated to professional development in the field of intellectual property management. Intellectual property has become vitally important in international commerce, and its effective management has become an essential skill in business. The mission of IPM*i* is therefore to encourage the development of intellectual property management theory and practice, and to provide recognition for those who, through educational and professional development, have attained professional status. Members prepare for the Candidate and Certified Intellectual Property Manager examinations through a self-study program. A detailed bibliography is available to Members to guide their study, as well as a directory to educational sources.

Membership in IPM*i*

There are three levels of membership available in IPM*i*:

- Member
- Intellectual Property Associate
- Certified Intellectual Property Manager (CIPM)

Members will be accepted upon furnishing evidence of the following qualifications:

- Graduation from a four-year college program
- A minimum of two years' professional experience in a field related to intellectual property
- Personal recommendation from three persons familiar with the applicant's reputation and business experience, one of whom should be the applicant's business or professional supervisor
- The college degree requirement can be waived in lieu of eight years of relevant professional experience

An Associate will have fulfilled the requirements for Member and satisfied these additional requirements:

- The successful completion of an examination, which will cover the following:

    —Ethics

    —Basic intellectual property legal concepts

    —Basic accounting principles

    —Basic principles of finance and economics

An Associate will receive the CIPM designation upon successful completion of an examination, which will cover the full subject matter of the IPM*i* syllabus. A Certified Intellectual Property Manager is entitled to use the CIPM designation. A Certified Intellectual Property Manager is required to maintain proficiency by participating in continuing education programs.

Dues

The annual dues for the various levels of membership in IPM*i* are as follows:

- Member—$125.00

- Intellectual Property Associate—$175.00
- Certified Intellectual Property Manager—$250.00

Fees

Membership application—$50.00

Associate examination—$100.00

CIPM examination—$200.00

IPM*i* awards the Certified Intellectual Property Manager (CIPM) designation to those members who are qualified by their education and business experience, and who successfully pass a rigorous, written, multidisciplinary examination to demonstrate their professionalism. Those who achieve their CIPM designation will be required to maintain their competence by satisfying continuing education requirements. An intermediate step in the process allows a member to attain the Associate designation. By maintaining rigorous professional standards and an examination program, the objective of IPM*i* is to make the CIPM designation a symbol of professional excellence in intellectual property management. CIPM designees will therefore enjoy a recognized position for having attained this professional standard, and for the possession of a unique multidisciplined skill set. The CIPM designation will be a worthwhile adjunct to other professional designations.

Program of Study

The following is an outline of a course of study, which would prepare a Member for the successful completion of the examinations for Associate and for CIPM. In each major section is provided a bibliography of reference materials and sources of educational programs, which are known at this time. IPM*i* does not provide its own educational programs. They recognize that IPM*i* Members will come from many professional disciplines and may already possess extensive background in one or more of these subject areas. The preparation for the Candidate and CIPM examinations is, therefore, one of self-study, and Members are provided with information regarding sources of written materials and educational programs to assist them in accomplishing their self-directed programs. In most cases, Members will find it necessary to utilize both sources because some intellectual property subjects are only available in seminar, workshop, or formal education settings.

When the term "IP" is used in the following, it is intended to encompass patented and unpatented technology, trade secrets, trademarks, copyrights, software, and Internet assets and related management systems.

The program of study for attaining the Associate membership status is outlined as follows:

Intellectual Property Associate

Ethics:

- Ethical responsibilities as corporate officer and/or advisor
  - —Conflicts of interest
  - —Disclosure
  - —Fiduciary responsibility

Legal:

- Establishing IP rights
  - —Federal
  - —State
  - —Common law
  - —Patents
  - —Trademarks
  - —Copyright
  - —Trade secret
  - —Protection geography
  - —Patent or not
  - —Patent vs. trade secret
  - —Patent and trademark protection synergies

Accounting/Taxation:

- Basic accounting principles
  - —Accounting theory
  - —Financial reporting

—Business statistics

—Cost accounting

—Balance sheet

—Income statement

—Sources and uses of funds

Finance/Economics:

- Sources of capital
- Basic economic principles

In order to complete the program and attain the CIPM designation, the following areas of knowledge and competence are regarded:

Certified Intellectual Property Manager

The Business of Intellectual Property:

- IP department organization/operation
- Managing the "business" side of an R&D program
- IP ownership issues (employee vs. company)
- IP creation and nurturing

—Harvesting inventions

—Invention disclosures

—Processing invention disclosures

—Inventor compensation

—Corporate incentive plans

—Patent solicitation

—Technical publications

—Patent marking

—Outside submissions

—Ownership of IP (patents, trade secrets, software)

—Management of employee inventions

—Employment/invention agreements vs. termination agreements/exit interviews

—Lab notebook keeping

—To patent or to padlock

- IP audits and due diligence
- Allocating company resources (evaluate the direction of the R&D budget)
- Understanding the objectives of management/stockholders
- Managing and enforcing in-house IP protection practices
- Participating in mergers and acquisition teams
- The IP holding company
- Developing corporate identity program
- Assisting business managers on strategy/forecasts
- Ensuring nonduplication of IP development
- Evaluating R&D resource allocation
- Acting as clearinghouse for outside IP searches
- Managing IP exploitation program

—Licensing

—Joint ventures and alliances

—Partnerships

—Strategy development

—Co-branding

—Franchising

—Distribution relationships

—Charitable donations

—Joint research and development

- Managing the interface with in-out house tax people

Communications:

- Communicating IP matters to management
- Communicating IP practices to inventors, R&D personnel, marketing, and advertising
- Social/business customs around the world

Ethics:

- (Covered in Associate program)

Marketing:

- Using/directing market research
- Making forecasts
- Interface with marketing/advertising (in-house and out)
- Understanding of price-volume-profit relationships
- Competitive intelligence
- Interface of markets and R&D programs

Human Resources:

- Searching, hiring, and retaining IP staff
- Training IP staff

Information Technology for IP Management:

- Database management of IP portfolio
- Dissemination of IP resources within the organization
- Using spreadsheets for analysis of IP financial performance
- Using Internet resources

Negotiating:

- Develop negotiating skills
  —For transactions

—In litigation

—Internal discussions

Legal:

- Basics (Covered in Associate program)
- Legal remedies

  —Injunctions

  —Damages
- Managing IP litigation and understanding its economics

  —Sword rattling to full-court press

  —Estimating litigation costs

  —Using outsiders

  —Working with experts

  —Controlling litigation costs

  —Foreign IP practice

  —IP remedies for damage (case law)

  —As the defendant

  —As the plaintiff

  —Estimating litigation success

  —Litigation alternatives
- Economics of various forms of protection
- Legal aspects of licensing, joint ventures, co-branding, etc.

  —International issues

  —Bankruptcy issues

  —Antitrust issues (Hart-Scott-Rodino and Dept. of Justice/FTC guidelines)
- Legal aspects of mergers and acquisitions

  —Due diligence

  —Tax implications

Accounting/Taxation:

- Basics (Covered in Associate program)
- Using public financial information
    - —Market intelligence
    - —Adversary investigation
    - —Searching for partners
    - —Evaluating licensee strength
    - —Finding and analyzing potential infringers
- Understanding international accounting standards
- Assisting in due diligence and accounting audit services
- Understanding accounting issues in licensing, joint ventures, co-branding, etc.
- Understanding the issues in IP taxation
    - —Transaction-related
        - Capital gains
        - IP development agreements
        - Taxation of royalties
        - Withholding
    - —International issues
    - —Transfer pricing
    - —Ad valorem taxation
    - —State tax issues

Finance/Economics/Valuation:

- Principles of finance
    - —Sources of capital
    - —Role of IP
    - —Mathematics of investment
    - —Financial markets

—Pricing products/services

—Financial statement analysis

—Capital budgeting

- Principles of economics

    —Managerial economics

- Economic evaluation of the forms of exploitation

    —Sale

    —Purchase

    —Licensing

    —Joint ventures and alliances

    —Swaps

    —Portfolio licensing

    —Searching for infringers

- Analyzing and quantifying IP "rent" in all of its forms
- Preparation and evaluation of business plans and financing alternatives
- Preparation of prospectus and offering materials
- Interfacing with financial institutions
- Acting as intermediary in licensing, sale, purchase, and joint ventures
- Knowing availability of exogenous sources of information
- Evaluating the effect of market research
- Understanding consumer/buyer behavior
- Knowledge of forecasting theory and available tools
- International business differences
- Banking and currency
- IP valuation/theory and practice

    —Premise of value

    —Value relationships

    —Cost, market, and income approaches

    —Discounted cash flow techniques

—IP assets and the business enterprise

—Monetary, tangible, and intangible assets

—IP assets as a portfolio

—Relative risks

- IP royalties/theory and practice

  —Sources of market data

  —Investment/rate of return techniques

  —Other analytical techniques

- Quantifying damages

  —Trademarks

  —Patents

  —Noninfringement business damages

International Issues:

- Doing business worldwide

  —Differences in culture and language

  —Essential differences in intellectual property law

  —Essential differences in taxes

  —Essential differences in accounting practices

  —The nature of worldwide markets for intellectual property

Industry Practices:

- The role and importance of intellectual property in various primary industries

  —Relative importance of various intellectual property types from industry to industry

  —The relative value, economic life, and risk of intellectual property from industry to industry

  —Essential financial reporting and taxation issues by industry

Intellectual Capital:

- The nature of intellectual capital
  - —Intellectual capital vs. intellectual property
  - —The importance of intellectual capital within a business enterprise
  - —The management of intellectual capital
  - —The quantification of intellectual capital

Career Paths for IPM*i* Members

The Institute foresees many career paths for professionals in intellectual property. Provided here is a list of career opportunities in several industries.

Corporation:

- Management of IP holding company
- Corporate staff including M&A groups
- Middle management positions at division, subsidiary, or business-unit level
- Corporate counsel

Law Firm:

- IP attorney (providing enhanced business services)
- IP attorney—litigation
- Clients without in-house capability

Consulting Firm:

- Consultant providing valuation, damages, strategic, or exploitation counsel
- Consultant supporting due diligence or auditing efforts
- Consultant acting as intermediary or broker
- Consultant providing IP management training

Accounting Firm:

- Consulting activities, as above
- Auditing of IP-intensive clients
- Due diligence of IP-intensive targets

Banking, Venture Capital, Intermediary:

- Part of a "deal-making" or due diligence team
- Financing/credit administration to IP-intensive clients

Government:

- Patent and Trademark Office management
- Taxation (federal/state)
- International trade
- Non-lawyer consultant to middle market

Advisory Council

The professional affairs of IPM*i* are implemented by the Advisory Council, which serves a keystone role in the organization. The members of the Advisory Council are recognized professionals in various fields related to the management of intellectual property. The experience of these professionals is key to the development of IPM*i* professional standards and the requirements associated with the CIPM designation. The Advisory Council members have primary input to the CIPM curriculum and examination standards. Ultimately, the Advisory Council will comprise approximately 15 members. At present the Advisory Council members are:

Allan Feldman, President, Leveraged Marketing Corporation of America

Allan Feldman founded Leveraged Marketing Corporation of America in 1983. LMCA is a full-service licensing firm working exclusively in the area of global brand licensing for major corporate clients. Mr. Feldman has more than 25 years' experience in marketing and management, including over 18 years in the area of brand extension licensing. He earned a master of business administration degree from the University of Chicago, where he was selected an International Fellow, and studied at the University of Louvain in Belgium and the London School of Economics. He also holds a bachelor of science degree from Roosevelt University. He is a member of the American Management Association, American Marketing Association, Licensing Executives Society, and the International Trademark Association.

Heinz Goddar, Partner, Boehmert & Boehmert, Germany

Dr. Goddar is a German and European patent attorney who is a partner with Boehmert & Boehmert and Forrester & Boehmert, having offices in Alicante, Berlin, Bremen, Düsseldorf, Frankfurt, Kiel, Leipzig, Munich, and Potsdam. He is responsible for international patent and licensing matters, including litigation and arbitration, with a special interest in the European Union. Dr. Goddar is an associate judge at the Senate for Patent Attorney Matters at the German Federal Supreme Court and lectures on the topic of patent and licensing law at the University of Bremen, Germany. His technical background, as well as his Ph.D. degree, is concentrated in the area of physics and physical chemistry. Prior to his career as a patent attorney, Dr. Goddar was an assistant professor in the Physical Chemical Department at the University of Mainz, Germany. Dr. Goddar is president of LES International and is immediate past president of LES Germany.

Robert Goldscheider, Chairman, International Licensing Network, Ltd.

Mr. Goldscheider, an attorney, is a well-known specialist on the commercial and legal aspects of technology transfer, has acted as a consultant to many multinational clients, and has served as an expert witness in litigation relative to intellectual property matters. He is the author of numerous IP publications and an active lecturer.

Theodoros K. Grigoriou, President, Honeywell Technologies, Inc.

Honeywell is an advanced technology and manufacturing company involved in the aerospace, automotive products, chemicals, fibers, plastics, and advanced materials industries. As president of Honeywell Technologies, Inc., Mr. Grigoriou is responsible for the management and exploitation of a large portfolio of the company's technology assets.

Edward J. Hendrick Jr., Vice President, Business Development and Technology Commercialization, SAIC

Ed Hendrick is vice president of business development and technology commercialization at Science Applications International Corporation (SAIC), a high-technology research and engineering firm serving both the government and commercial sectors. He began his R&D career at Bell Laboratories in 1966 and served in many management positions, including the team responsible for the design and development of the first USA-made digital switch. He held subsequent positions at Bellcore and U.S. West, where he was involved in licensing, strategic technology planning, and intellectual property optimization. Mr. Hendrick is a graduate of St. Joseph's University and has his MBA from Fairleigh Dickinson University. He also earned both juris doctor and master in intellectual property degrees from Franklin Pierce Law Center.

Karl F. Jorda, Director, Germeshausen Center for the Law of Innovation and Entrepreneurship, Franklin Pierce Law Center

Karl F. Jorda is director of the Germeshausen Center for the Law of Innovation and Entrepreneurship and the David Rines Professor of Intellectual Property Law at Franklin Pierce Law Center. He is also an adjunct professor at the Fletcher School of Law and Diplomacy, Tufts University. His courses cover a wide spectrum of IP law and business subjects. The 1996 recipient of the Jefferson Medal of the New Jersey Intellectual Property Law Association, Professor Jorda also serves as a consultant to corporations, governments, the World Intellectual Property Organization, and other international agencies throughout the world. He is a tireless world traveler on behalf of Franklin Pierce Law Center and the cause of intellectual property education. Prior to joining the Law Center, he was chief IP counsel at Ciba-Geigy (now Novartis) for 26 years. Professor Jorda received his undergraduate degree from the University of Great Falls, and master's and law degrees from Notre Dame University. He is admitted to the bars of Illinois, Indiana, and New York, as well as to practice before the U.S. Patent and Trademark Office.

Sam Khoury, President, INAVISIS, Inc.

Mr. Khoury holds a doctorate in chemical engineering and a master's degree in business administration. He served as the senior intangible asset appraiser for the Dow Chemical Company and has now founded INAVISIS, Inc., a firm dedicated to intellectual property consulting. He has been instrumental in the seminal development of IP management systems.

Willy Manfroy, Bornival LLC.

Mr. Manfroy was employed at the Dow Chemical Company for 25 years and subsequently was director of corporate development at Eastman Chemical Company, where his responsibilities included acquisitions, divestitures, and licensing. He is a past president of the Licensing Executives Society (U.S.A. and Canada). In his current consulting position, he serves a worldwide clientele on business matters concerning the management and exploitation of intellectual property.

Daniel M. McGavock, President and Managing Director, InteCap, Inc.

Dan McGavock is president and managing director of InteCap, Inc. InteCap, formerly IPC Group, is a national consulting firm with a professional staff of more than 180 professionals operating out of 10 offices throughout the United States. InteCap advises clients on economic, valuation, and strategy issues related to intellectual property and complex commercial disputes. Mr. McGavock was an originator of his firm's Intellectual Property Quality Management (IPQMSM) practice, which is dedicated to helping companies maximize shareholder value through the strategic management of intellectual property. A graduate of Indiana University in accounting, he is a CPA in the state of Illinois and a member of the American Institute of Certified Public Accountants and is an arbitrator of the American Arbitration Association. He has been very active in the Licensing Executives Society (USA and Canada) and the International Trademark Association. He has provided expert testimony in federal court, state court, tax court, and in arbitration proceedings on the subjects of economic damages and intellectual property valuation issues.

Emmett J. Murtha, President, Fairfield Resources International, Inc.

Emmett Murtha was employed by IBM Corporation for 35 years, during which time he was director of IBM's licensing program, was responsible for IBM's worldwide licensing policies and practices, and also served as IBM's director of business development, with responsibility to leverage IBM's intellectual property. A founding principal of FRI, he advises multinational clients on a wide variety of IP matters. Mr. Murtha is president of the Licensing Executives Society (U.S.A. and Canada).

Susan M. Richey, Professor of Law, Franklin Pierce Law Center

After receiving her law degree from the University of Maryland, Ms. Richey served as a staff law clerk for the U.S. Court of Appeals for the Fourth Circuit, was a principal with the firm of Riordan & McKinzie, and had an extensive practice in trademark and copyright litigation. She heads the Trademark and Copyright curriculum at Franklin Pierce Law Center.

Patrick H. Sullivan, Partner, The ICM Group

Dr. Sullivan is a founding partner of The ICM Group, a consulting company which is focused on extracting value from intellectual capital. He is also co-founder and convener of the prestigious ICM Gathering, comprising of managers of intellectual capital for large international companies who meet to exchange information on new and innovative management techniques. Dr. Sullivan has an undergraduate degree in engineering, a master's degree in research and development management, and a doctorate in business administration. He worked as an engineer on the launch team of the Saturn/Apollo project and has been a principle consultant at SRI International, where he managed the firm's general consulting practice in Europe. Dr. Sullivan is a Fellow of the American Council on Education and is a member of the American Bar Association Intellectual Property Section, the Licensing Executives Society, and the World Intellectual Property Trade Forum.

Membership Application

Visit the web site of IPM*i* and download its membership application—*www.ipinstitute.com.*

INDEX